Understanding Maps

Understanding Maps

A systematic history
of their use and development

A. G. HODGKISS

Senior Experimental Officer, Department of Geography
University of Liverpool

DAWSON

First published in 1981

Wm Dawson & Son Ltd, Cannon House,
Folkestone, Kent, England

British Library Cataloguing in Publication Data

Hodgkiss, Alan Geoffrey
  Understanding maps.
  1. Cartography – History
  I. Title
  526′.09   GA201
  ISBN 0-7129-0940-0

Printed and bound in Great Britain
by Mackays of Chatham

CONTENTS

Illustrations   7

Acknowledgements   10

1 Maps as a medium of communication   11
2 Fundamentals of mapmaking: (i) The development of
   a cartographic language   25
3 Fundamentals of mapmaking: (ii) The development
   of a cartographic vocabulary   39
4 Making the map   52
5 The evolving world map   71
6 Regional maps   86
7 Nautical charts   103
8 Route maps   119
9 Town plans and views   133
10 Thematic maps   154
11 Official mapmaking today   169
12 Modern commercial and private cartography   184

Select bibliography   199
Index   204

'I am told that there are people who do not care for maps, and find it hard to believe. The names, the shapes . . . the courses of the roads and rivers . . . are an inexhaustible fund of interest for any man with eyes to see or two-penceworth of imagination to understand with.'

R. L. Stevenson in *Treasure Island*

ILLUSTRATIONS

page

1 Maps in miniature: philatelic cartography 12
2 Mural map designed by F. Macdonald Gill from the 'Queen Mary' 13
3 Contrasting detail in two maps of Jutland from Ortelius's *Theatrum Orbis Terrarum* (1570) 14
4 Nautical chart of the Caribbean area by Jean Rotz (1542) 16
5 Playing cards with maps by J. Hoffmann and W. Redmayne 17
6 Dissected world map by Laurie & Whittle (1794) 18
7 Woodcut map of the island of Utopia from Sir Thomas More's *Utopia* (1518) 19
8 Political cartoon map, *The Evil Genius of Europe* (1859) 20
9 Stages of the mapmaking communication process (after Woodward) 22
10 Sources of 'noise' in cartographic communication 23
11 County map of Cornwall from Saxton's *Atlas of England and Wales* (1579) 24
12 Cartouche from Coronelli's China (c 1600) 26
13 *Cambridgeshire* by John Speed (1611) illustrating representation of scale 28
14 Africa from Ramusio's *Delle Navigationi e Viaggi* (1556) showing effect of orientation to the south 29
15 Nicholas Sanson's map of Europe (1668) based on Prime Meridian of Ferro 31
16 Four types of reference system: Cartesian Co-ordinates; Range and Township; Centuriation; UTM 32
17 Three types of perspective projection 34
18 Michael Mercator's map of the Americas (1606) on a stereographic projection 36
19 Johan Ruysch's world map on a conical projection in Ptolemy's *Geographia* (Rome, 1508) 37
20 Varied ways in which mapmakers have attempted to provide a graphic impression of relief 40
21 Map of the North African region from Ptolemy's *Geographia* (Rome, 1490) displaying techniques of illustrating natural phenomena 41
22 Detail from Eduard Imhof's wall map of Kanton Graubünden, 1:100,000 42
23 Ways in which cartographers have differentiated sea from land 43
24 The church as a symbol for settlement in the sixteenth, seventeenth and eighteenth centuries 45
25 Taprobana by Sebastian Münster from his edition of Ptolemy's *Geographia* (Basle, 1540) 46
26 Anthony Jenkinson's map of Russia from Ortelius's *Theatrum Orbis Terrarum* (1570) 47
27 Examples of calligraphy used on maps from the fifteenth to the nineteenth century 48
28 Münster's map of the New World (1540) showing lettering executed by inserting metal type into the wood block 49
29 Cartouche from John Mitchell's *A Map of the British Colonies in North America* (1755) 50
30 Triangulation diagram from William Yates's map of Lancashire (1786) 54
31 Detail from *Carte Géometrique de la France* produced between 1744 and 1789 by the Cassini family 56
32 Satellite image of Chesapeake Bay produced by LANDSAT 1 satellite in 1977 57
33 Infra-red image taken by the NOAA-2 satellite 59
34 Detail from world chart by Pierre Desceliers (1487–1553) 60
35 Detail from chart of north-western Europe by Homen (1569) 62
36 Detail from Sheldon tapestry map (c 1588) 63
37 Woodcut map of Europe from Bordone's *Isolario* (1528) 64

38 Examples of tools used in copper engraving   66
39 Isoline map prepared by computer and graph plotter showing areas in which a sample of Portsmouth dwellers would prefer to live   68
40 Babylonian world map (*c* 500 BC)   71
41 Reconstruction of Eratosthenes' reference grid   73
42 Detail from a reconstruction of the al-Idrisi world map (twelfth century) showing the British Isles   74
43 Arabic world map by al-Idrisi from the *Book of Roger*   75
44 Diagram showing the principle of the T-O or T in O map   75
45 Early fourteenth-century T-O map by Brunetto Latini   76
46 Climatic zone map by Macrobius (eleventh century)   76
47 Cottonian or Anglo-Saxon map (*c* 1000)   77
48 Hereford Map, made *c* 1300 by Richard of Haldingham   79
49 Woodcut world map from Ptolemy's *Geographia* (Ulm, 1482)   81
50 Detail from a facsimile of Juan de la Cosa's world map (1500)   82
51 World map on a conical projection by G. M. Contarini (1506)   83
52 Diego Ribeiro's world map (1529)   84
53 The world outline as seen by la Cosa, Ribeiro, Waldseemüller and Mercator   85
54 Clay tablet map (*c* 3800 BC) showing part of northern Mesopotamia   86
55 Roman map of garrisons in the Nile valley from *Notitia Dignitatum* (fifth century AD)   87
56 Detail from Matthew Paris's map of the Holy Land (*c* 1250)   89
57 Detail showing the Thames estuary from the Gough or Bodleian Map (*c* 1360)   90
58 Sebastian Münster's woodcut map of Franconia (1540) showing contemporary cartographic conventions and techniques   91
59 'Indiae Tabula Moderna' from the Strasburg edition of Ptolemy's *Geographia* (1525)   92
60 Detail from G. B. Ramusio's map of Brazil in *Delle Navigationi e Viaggi* (1556)   93
61 Map of Calabria from Ortelius's *Theatrum Orbis Terrarum* (1606 ed)   94
62 Devonshire from Saxton's *Atlas of England and Wales* (1579)   96
63 Leicester and Rutland by William Smith (1602–3)   97
64 Berkshire from Speed's *Theatre of the Empire of Great Britaine* (1611)   98

65 Detail from Donn's prize-winning map of Devonshire (1765)   99
66 John White's map of Virginia (1590)   100
67 Detail from a facsimile of John Mitchell's *Map of the British and French Dominions in North America* (1755)   101
68 Captain John Smith's map of Virginia (1612)   102
69 Detail from a facsimile of the *Catalan Atlas* (1375)   105
70 Britain and north-western Europe from Bordone's *Isolario* (1547 ed)   106
71 Chart of southern Europe and north Africa by Bartolomeo Olives (1559)   107
72 Chart of the Cornish coast from L. J. Waghenaer's *De Spieghel der Zeevaert* (1584)   110
73 Chart of the Dee estuary from Collins's *Great Britain's Coasting Pilot* (1693)   111
74 Murdoch Mackenzie's chart of the north-west coast of Orkney (1742)   113
75 Chart of the harbour of Rhode Island and Narraganset Bay (1776) from *Atlantic Neptune* by J. F. W. Des Barres   114
76 List of symbols used by Captain H. M. Denham in his surveys of the Mersey and Dee (1840)   116
77 Detail from a modern chart of Liverpool Bay prepared by the Mersey Docks and Harbour Company   117
78 Detail from the Peutinger Table   120
79 Triangular distance table from Morden and Cox's *Magna Britannia* (1738)   122
80 Strip map from Ogilby's *Britannia* (1675)   123
81 Strip map from *Taylor and Skinner's Survey and Maps of the Roads of North Britain or Scotland* (1776)   124
82 Detail from Bradshaw's *Map of Canals, Navigable Rivers, Railways etc. in the Southern Counties of England* (*c* 1830)   126
83 Plan and section of the Grand Junction Railway from *Cornish's Guide and Companion to the Grand Junction and the Liverpool and Manchester Railways* (3rd ed, 1838)   128
84 Map of the Adirondacks from *The Englishman's Guide Book to the United States and Canada* (1884 ed)   130
85 Perspective view of Venice from Bordone's *Isolario* (1547 ed)   134
86 Profile of Nuremberg from Schedel's *Nuremberg Chronicle* (1493)   135
87 Detail from Münster's perspective view of Constantinople from *Cosmographiae Universalis* (1550 ed)   136

88 Elevation of Nuremberg (1552) by Hans Lautensack 137

89 Mexico City from Ramusio's *Delle Navigationi e Viaggi* (1556) 138

90 London from Braun and Hogenberg's *Civitates Orbis Terrarum* (1572–1618) 140

91 Inset plan of Denbigh from Speed's *Denbighshire* (1611) 141

92 Pictorial plan of Gubbio from *Novum Italiae Theatrum*, the 1724 edition of Blaeu's town-books of Italy 142

93 View, plan and location map of Dunkirk by S. de Beaurain (*c* 1760) 143

94 Detail from Turgot's plan of Paris (1734–39) 144

95 Detail from Horwood's plan of London (1792–9) 146

96 Plate from Camille N. Dry's pictorial plan of St Louis (1875) 148

97 Banks & Co.'s balloon view of London from Hampstead (1851) 149

98 Detail from Ordnance Survey 1:528 plan of central Nantwich (1878) 150

99 Detail from Hermann Bollmann's plan of Hamburg 151

100 Detail from Bollmann's view of Central Manhatten 152

101 Qualitative map based on linear data showing Continental Trailways bus system 155

102 Use of *isobaths* by Cruquius in his map of the Merwede river bed (1729) 157

103 Isogonic map (1701) by Edmond Halley entitled *A New and Correct Chart showing the Variation of the Compass in the Western and Southern Oceans* 158

104 Christopher Packe's *A New Philosophico-Chorographical Chart of East Kent* (1743) 160

105 Ethnographic map from Berghaus's *Physikalischer Atlas* (1850) 162

106 Map of Ireland by Henry D. Harness (1837) using innovatory techniques of statistical presentation 163

107 Map of social and ecclesiastical conditions in Liverpool (1858) by the Rev. A. Hume 165

108 Detail from a facsimile of Thomas Milne's land utilisation map of the London region (1800) 166

109 Detail from Ordnance Survey First Series One Inch to One Mile, Sheet 11 (1810) 171

110 Detail from *Carte Topographique de la Suisse, Carte Dufour*, 1:100,000, showing part of the Rhône valley 173

111 Detail from 1:50,00 map of Hong Kong 176

112 Detail from the New Zealand Topographical Map, 1:25,000, NZMS Taranaki, Sheet 7, 1st edition, 1970, showing Mt Egmont 177

113 Directorate of Overseas Survey Map DOS 310, Deception Island, Edition 1 178

114 Detail from the Peyto Glacier map prepared by the Inland Waters Directorate of the Department of the Environment, Canada 180

115 Detail from United States Geological Survey 1:125,000, Oregon-Washington, Mount Hood Quadrangle 181

116 Detail from United States Geological Survey, 1:24,000, Philadelphia Quadrangle, 1967 182

117 Detail from the *Commuter's Map of 60 Miles around London* (original edition published by John Swan & Co., London) 185

118 Detail from plan of Milton Keynes prepared by David L. Fryer & Co. 186

119 TWA press advertisement of the 'Atlantic River' 189

120 Decorative map of the wine-producing district of Boberg in the Republic of South Africa prepared by the Janice Ashby Design Studio 190

121 The 'map model' technique used in a poster map for the Fyffes Group 192

122 Propaganda map of Cyprus (1965) 193

123 Poster map of British Festivals and Customs used to further the National Savings Campaign 195

124 Pictorial map of Surrey by the artist, Montague Webb 196

125 Heraldic map of Devonshire (1972) by T. Alan Keith-Hill 197

126 Panoramic map painted for the 1976 Winter Olympics by Heinrich Berann 198

ACKNOWLEDGEMENTS

I would like to express my appreciation of the generous help I have received from many sources. In particular I am indebted to Dr J. B. Harley for his unfailingly constructive advice and pertinent comments which have improved my work immeasurably. Many librarians and map curators have given me assistance but I owe especial thanks to Michael Perkin and John Clegg who have been unfailingly patient and helpful in making available the maps and early atlases in the Sydney Jones Library, University of Liverpool. Elizabeth Arkell of the Bodleian Library, Oxford and Sarah Tyacke of the British Library have responded with exemplary efficiency and courtesy to my numerous enquiries. The photography of maps in early atlases is no easy task and I owe grateful thanks to Douglas Birch, Harry Taylor, Ian Qualtrough, Arthur West, Kenneth Garfield and Christopher Lewis for a vast amount of photographic assistance. My thanks are also due to the many organisations and individuals who have generously supplied me with information, copies of their maps or permitted their work to be reproduced. Heinrich Berann, Hermann Bollmann, Carlos Sanz, Eduard Imhof and Alan Keith-Hill have been exceptionally kind and helpful. My family have faced with fortitude a home littered with maps and atlases of every kind during the writing of this book and I am appreciative of their forebearance. Finaly I thank Meg Davies for her efficient copy-editing of my manuscript and my editor, Robert Seal, for his constant interest and courteous assistance.

A. G. Hodgkiss
Liverpool, May 1980

1 MAPS AS A MEDIUM OF COMMUNICATION

In 1703 William Alingham wrote that 'Maps are . . . an Invention of such vast Use to Mankind, that there is scarce anything for which the World is more indebted to the Studies of Ingenious Men, than this of describing Maps.' Since Alingham's time the nature and uses of maps have changed dramatically but his words, nevertheless, remain apposite today. Maps, in his day, reached only a limited public of literate, educated persons; today they are an integral part of daily life. It is difficult to avoid being confronted by at least one or two maps during the daily routine. Perusing the morning paper in the commuter train we are likely to see small black-and-white maps serving to locate and explain some significant contemporary event. At home in the evening similar maps face us on the television screen, as a feature of the television news. The current state of the weather is indicated in the press and on television with the aid of satellite photographs and maps which have been specially designed so that their meaning should be clear to the untrained map user.

If we stop to consider this brief confrontation with maps we find that, perhaps surprisingly, we have already been brought face to face with several facts about maps in general: they can be designed to meet a specific need or to illustrate a specialised topic (indeed it has been said that anything can be mapped, providing that data are available); they can be in colour or monochrome, depending on the media or the circumstances in which they appear; they can vary enormously in size; they can show varying amounts of detail; they can be designed for the understanding of different classes of users; they can be ephemeral (those of the television screen) or tangible and lasting.

If we look a little further into the nature and uses of maps it soon becomes apparent that an extraordinary diversity of maps is available to meet the requirements of an extensive range of map users. Maps can serve 'popular' needs such as those of walkers, climbers, motorists, yachtsmen or skiers. They can also serve as a scholarly tool of academic workers in numerous disciplines: historians, economists, sociologists, archaeologists, geologists and so on.

In today's society maps have a multi-faceted role which is the cumulative result of much thought and human endeavour over thousands of years, for the making of maps, albeit of a rudimentary nature, goes back so far as to predate written history.

WHAT IS A MAP?

Briefly, a map is a form of graphic communication designed to convey information about the environment. It provides a scaled-down view of reality, extending the observer's range of vision so that he sees before him a picture of a portion, perhaps a large portion or even the whole, of the earth's surface (or of some celestial body, for not all maps are earthbound). A major function of maps is to assist in the determination and understanding of geographical relationships and to this end maps are structured in a similar way to the territory they represent, with geographical phenomena located in their natural spatial relationships.

In the early days of mapmaking, maps were simple in content, doing little more than locate coastlines, rivers, hills and settlements. Their message could be understood, therefore, with little specialised training on the user's part. In the twentieth century, however, maps have become increasingly complex – due partly to improved methods and technology and partly to the wealth of information to be mapped – and the uninitiated user may experience difficulty in appreciating that the arrangement of lines, shapes, symbols, colours and names which constitutes a map can be a microcosm of reality.

THE LIMITATIONS OF MAPS

One of the important things to understand about maps is that however great their degree of sophistication they are subject to unavoidable restrictions. One of these is

Fig 1 **Maps can vary enormously in size. Philatelic cartography is mapmaking in miniature but maps on stamps serve a useful purpose in communicating information about small or newly-independent territories or in letting the world know about significant happenings in different countries. They can even be controversial e.g. the *13 paisa* Pakistan stamp describes Jammu and Kashmir as 'final status not yet determined' while the maps on Indian stamps show the territories as belonging to India.**

the incompatibility between the curved surface of the earth and the flat sheet of paper on which a map is drawn or printed: try peeling an orange and laying the peel flat to appreciate this difficulty! The curved surface cannot be transferred to a flat sheet without some distortion of areas, shapes, direction or distance. Although devices known as *map projections* have been developed in an attempt to solve the problem maps remain subject to a degree of distortion and no map can be perfect in this respect. The only way of achieving a really accurate representation of the earth is by means of a globe, but the globe has even greater limitations than the map owing to its restricted size, which allows only for a very

small-scale global presentation, and conversely to its bulk, which prohibits portability and confines the use of a globe to that of a teaching aid. Mapmakers have therefore persisted with the use of paper which is after all a functional medium – compact, light, reasonably strong and ideally suited to the printing process.

Maps are made in a proportional relationship to the territory they represent – the concept known as *scale*. The scale at which a map is drawn determines the amount and intricacy of detail to be included. The larger the scale, the greater can be the amount and complexity of the detail.

Another restriction is one of *selectivity* for, unlike aerial photographs which record everything seen by the airborne camera, maps are selective views and the compiler has to decide on those features which he would like to include and those which must be omitted. Any map is therefore a *generalised*, subjective view of reality. It communicates the view of its maker. It is also a *simplified* view, for the exigencies of a greatly reduced scale do not allow every bend of a river or road, or all the tiny bays of a coastline, to be reproduced.

Fig 2 At the opposite extreme to philatelic cartography is this large mural map which was designed by F. Macdonald Gill and was a prominent feature of C Deck Restaurant on the 'Queen Mary'.
Courtesy: Cunard Line and University of Liverpool

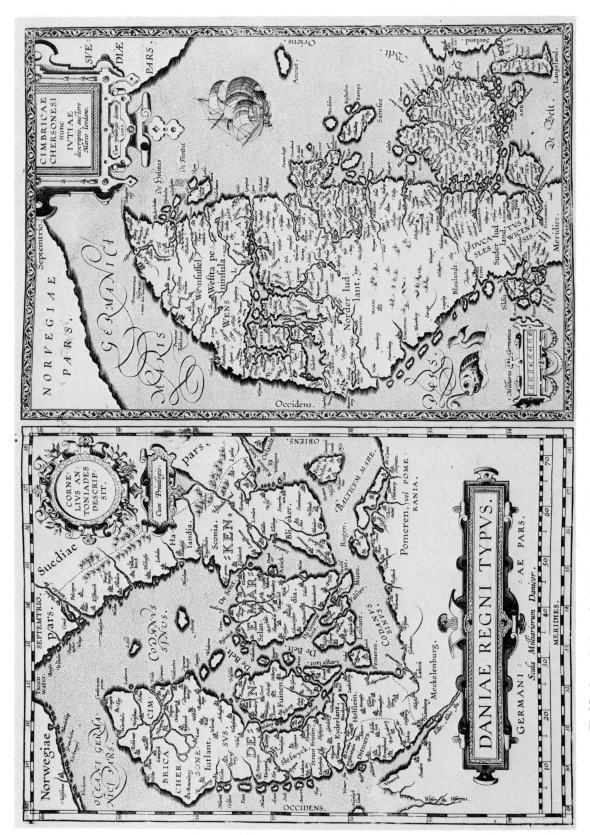

Fig 3 In the early days of printed maps little skill was required to interpret the maps' message for information was largely confined to coastlines, rivers, hills and places. What, however, was the user to make of this pair of maps which appeared side by side in Ortelius's *Theatrum Orbis Terrarum* (1570)? The delineation of Jutland in the larger-scale map on the right bears little resemblance to that of the left-hand map and the islands are very different. Beautiful engraving and layout ensure effective communication but *meaningful* communication is hindered by the topographical inaccuracies.

Courtesy: University of Liverpool   RvI N.1.3

THE USES OF MAPS

That maps have an almost inexhaustible range of uses has been implied already. The most familiar 'popular' use is that of route finding. The ubiquitous map in this category is the motoring map, issued in millions all over the world by official mapmaking agencies, commercial establishments, motoring organisations and petrol companies. Automobile maps generally are strictly functional and, while they may provide some information about services *en route* and even suggest interesting features which might warrant a detour, they contain less geographical information than the *topographical* map, a general-purpose kind of map such as the 1:50,000 series of the Ordnance Survey and many mapmaking organisations. With the contemporary insistence on speed of travel and the continuing expansion of motorway networks there is a danger that maps will become superfluous to many motorists. Motoring could become a purely mechanical exercise, the driver encapsulated in his vehicle knowing nothing of the countryside through which he is passing, and experiencing none of the excitement and anticipation engendered by a carefully-planned journey in which maps have been studied to work out a route of maximum interest. In less frenzied leisure activities such as mountain-walking, maps remain an invaluable route-finding aid and source of information about the terrain. Not only do they add interest to the walk but they can also serve as an aid to safety, pinpointing the hazards of walking the hills – the marshes to be skirted, the steep crags to be treated with care, the streams to be forded. In this context map design is supremely important and there must be no possibility of misunderstanding the map's message.

The tourist and holidaymaker is well catered for by the mapmaking industry. Indeed, the prospective holidaymaker ignores maps at his peril for they can convey a good deal to him about the quality and suitability of his proposed holiday area. So many complaints that a hotel is two miles from the expected beach, fronts on to a busy main road, is affected by airport noise and subject to industrial pollution from nearby factories could be avoided if tourists took the trouble to check with a suitable map.

Maps can influence choice in a number of ways. That of holiday venues has been mentioned, but topographical maps can provide a useful correction or confirmation of the mental picture we have of a locality when we contemplate a change of residence, a move to a new post, or seek a congenial place for retirement. The value of maps as a persuader has been seized on by advertising promoters who have been quick to appreciate the possibilities of colourful, eye-catching maps as an adjunct to their product promotions. To some extent advertisers play on the susceptibility of people to the charms of a decorative map and also on the way in which the public is prone to accept the message of any map as infallible. While conventional maps are as honest and reliable as the mapmaker can make them, less scrupulous mapmakers, particularly in wartime, produce propaganda maps which by the manipulation of symbols and colours and the distortion of areas and distances are deliberately designed to mislead. Another variable which can be used to convey a totally wrong impression is that of selection. The London firm of designers known as 'Diagram' prepared a map illustrating German and Russian advances into Poland at the beginning of World War II. The map was later reproduced in a Chinese publication. In order to represent the Russians as the only aggressors, the Chinese simply left off the arrows which had been used to show German advances.

Mapmaking can also exercise a straightforward military role in wartime with maps a valuable aid in arriving at many strategic decisions. It is hardly surprising, therefore, that military considerations, usually defensive, have provided a stimulus for mapmaking in most countries. Less obvious is the part played by the military in furthering the understanding of maps. In World War II, for example, troops received some basic training in map reading, and many will retain memories of outdoor map reading sessions in which they endeavoured to relate features on the ground to those on the map.

An extensive group of maps is exclusively concerned with the needs of navigation. These are more properly termed 'charts' and those for use in navigating at sea are *nautical charts*. The earliest extant specimen dates from 1300. Nautical charts today convey a vast amount of information for navigators at every level, from the lone yachtsman to the navigating officer on a bulk carrier. Accuracy and up-to-date revision are of the greatest importance, particularly in the representation of hazards such as shoals, wrecks and dangerous reefs. Charts vary in scale from large-scale charts of harbours to small-scale charts for ocean use. In recent times a new type of chart has been developed for use in aviation. These *aeronautical charts* are small-scale, due to the speed at which a plane travels, and are strongly coloured with the emphasis on altitudes of hills, location of landmarks, airfields, radio beams and beacons.

Orienteering is a unique example of a sport in which success depends almost entirely on efficient map reading. It was developed in Scandinavia sixty years ago but in the past twenty years has spread through Europe and into North America, Australia, New Zealand and Japan.

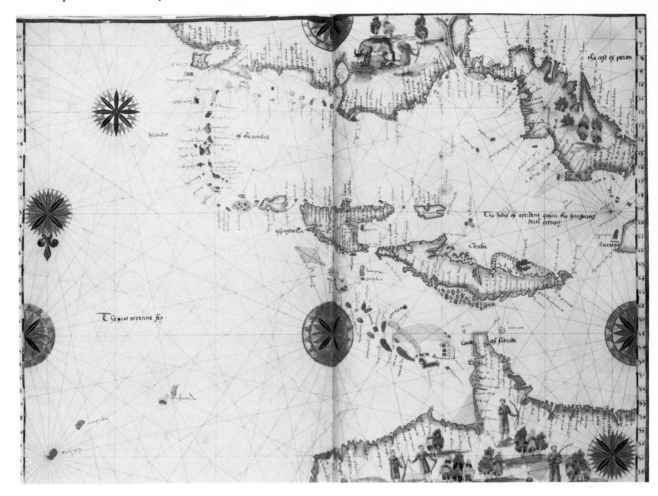

Fig 4 **A decorative nautical chart from the atlas of Jean Rotz, 1542. Rotz was one of the Dieppe school of mapmakers which produced some of the most artistic maps of any period. In this map of the Caribbean Rotz places south at the top of the map, a departure from convention which hinders communication to an observer who has grown to expect north at the top.**
Courtesy: **British Library Board   Royal MS, 20E 1X, ff23–24**

The sport involves cross-country navigation on foot using map and compass to reach specified check points. In the early days of the sport orienteers used standard topographical maps but as it developed these were found to be insufficiently detailed and often out-of-date. Nowadays, specially prepared orienteering maps are produced for the sport.

As a scholarly tool the advantages of maps seem obvious. They serve several functions in the educational field, their use as a teaching aid probably being the most widespread. Wall maps and globes remain popular in the classroom but in higher education greater reliance is placed on projected maps which are specially prepared by draughtsmen to illustrate concepts the lecturer wishes to communicate to his students. Maps are useful in other

disciplines as well as geography: the botanist may require maps showing the distribution of varied flora; the biologist, maps of bird migration; the historian, maps showing the political situation at different periods; the archaeologist, large-scale plans of ancient sites. These are just a few of the more obvious uses. It is no accident, however, that with its basic concern for spatial relationships geography has always been closely linked with mapmaking. Many early geographers, Mercator being the most famous example, were also mapmakers, and today geography departments in colleges and universities produce a wide range of maps and atlases associated with academic research. The cartographer's basic objective is to supply a compact, visual means of conveying information about the environment. His maps serve to record spatial phenomena at a much-reduced scale so that these phenomena can be clearly visualised for study purposes.

From the sixteenth century onwards, mapmakers have tried to enliven the study of geography by introducing an element of pleasure. In 1590 William Bowes, in a moment of inspiration, realised that the number of

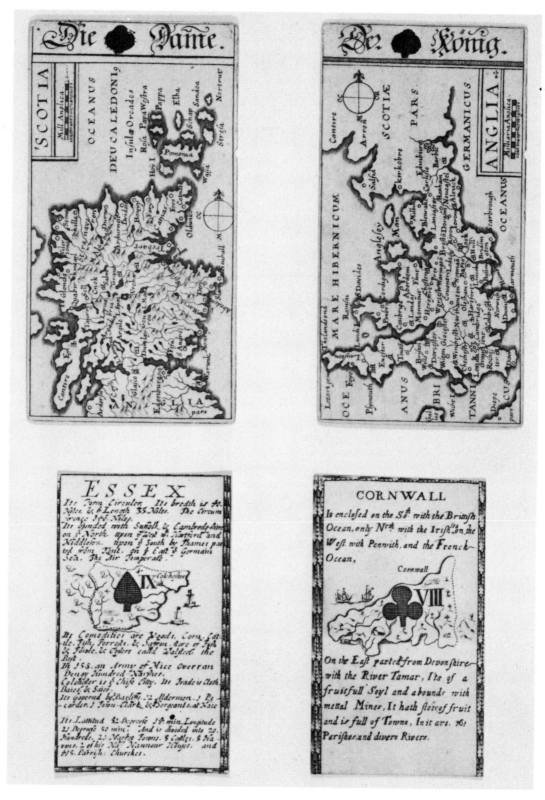

Fig 5 **Playing cards with maps serve to educate as well as entertain. The upper pair are by J. Hoffmann (Nuremberg, 1678), the lower by W. Redmayne (1711–12).**
Courtesy: **Worshipful Company of Makers of Playing Cards**

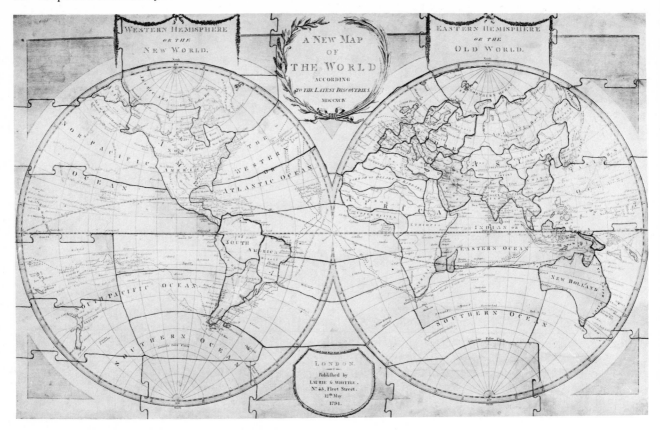

Fig 6 **Dissected maps have been a painless way of assimilating geographical information since the late-18th century. This dissected world map is by Laurie & Whittle, London, 1794.** *Courtesy*: **The Museum of London**

playing cards in a pack corresponded with the number of counties in England and Wales and produced a pack of cards, each of which bore a small county map. In 1676 Robert Morden issued a pack which had the distinction of being the first set of county maps to show roads. Other mapmakers issued their own packs, often giving a certain amount of statistical information about the territory illustrated.

Table games based on maps have been popular since the mid-seventeenth century. Many were of the race game type in which the player moved a set distance along a course according to the dictates of a dice or the spin of a teetotum. In 1652 the prolific French mapmaker, Pierre du Val, inventor of several geographical games, produced a draughtsboard in which the squares, instead of being alternately black or white, were either plain or contained tiny maps of the regions of France.

Dissected maps, now known as jigsaw puzzles, were introduced in the 1760s by John Spilsbury. In the early map jigsaws cutting took place along coastlines and national boundaries, a method which was helpful in assisting the child to remember both the shape and location of different countries.

It is also worth mentioning the use of maps in literature. Since the introduction of printing, many maps of fictional countries have been prepared, the earliest known example being that which described the imagined topography of Sir Thomas More's *Utopia* (1516). It is well known that Daniel Defoe made a map for *Gulliver's Travels* (1726). Robert Louis Stevenson, a confirmed map enthusiast, made a colourful map of an island in 1881 and subsequently based his famous novel *Treasure Island* on the map. Among the best-known contemporary fictional maps are those prepared to illustrate and publicise the novels of J. R. R. Tolkien. Humorous maps were popular during the late eighteenth and nineteenth centuries: *Geography Bewitched! or a droll Caricature Map of England and Wales* is one example. It shows England and Wales in the shape of a pipe-smoking, hard-drinking man riding on a fish. Such maps served merely to amuse and there was none of the satirical intent seen in the political cartoons of the late nineteenth century. The *Serio-Comic War Map for the year 1877* by Frederick Rose is an example of the latter. It is a map of Europe in which Russia, depicted as a large octopus, curls its tentacles in all directions to grasp at territories in different parts of Europe.

Fig 7 **Maps portraying the fictional lands of literature abound.
The illustration shows a woodcut map of the island of Utopia
from Sir Thomas More's** *Utopia* **(1518).**
Courtesy: **University College of St David's, Lampeter**

CLASSIFICATION OF MAPS

The quantity and diversity of maps produced today
necessitates some rational system of classification for
reference and study. Such a need has increased recently,
for maps throughout much of mapmaking history served
mainly to indicate 'what is where' and did not differ
greatly in basic characteristics. Early mapmaking can be
divided into the era of manuscript mapmaking and that
of the engraved or printed map. Although the earliest
European printed map appeared in 1472 the dividing
line can be set at 1500, the century between 1472 and
1570 being regarded as a transitional period. Early car-

tography can then be divided into the making of general
maps, nautical charts, town plans and globes, and early
printed maps can be grouped first by the map subject;
secondly by the name of the mapmaker, printer and
publisher; thirdly by the method of production – wood
block or engraved metal plate; fourthly by the date of
issue.

Modern maps require a rather different classification
but luckily there is again a broad natural division into
two groups: first, the *special-purpose thematic map*, and
secondly, the *general-purpose reference map*. Special-
purpose maps are not new – the Babylonians and Egyptians in the ancient world made large-scale maps for
taxation purposes (*cadastral maps*), the Romans made
route maps to assist in administration, medieval *mappæ
mundi* served to communicate theological beliefs, and
from the seventeenth century onwards military maps of
battle campaigns, siege plans and maps of fortifications
were common. The true thematic map as we know it,
however, developed after 1701, the year in which
Edmond Halley produced his map of *isogonic lines*, i.e.
lines passing through points of equal compass declination. In the late eighteenth century and throughout the
nineteenth there was a burgeoning of scientific and
social data relating to the human environment and it was
logical that mapmakers and geographers should begin to
examine the nature of phenomena at different locations
and investigate the size, quantity or value of such features. Their researches resulted in a prolific output of
maps on a variety of themes.

Reference maps serve as a means of looking up the
location of geographical features. On them no emphasis
is laid on any one feature at the expense of others, each
item being given a roughly similar visual stress. Reference maps are usually produced by official mapping
agencies or the larger commercial firms. They may be
classified regionally and according to their scale, which
can vary from the very small scale of a general world map
to the much larger scales of national map series. Thematic maps with their single function are designed to
emphasise a specific theme or concept. Many are published in sheet map form, others appear in atlases, textbooks and journals. They can be separated into two
main groups – *qualitative* maps which show merely the
nature and location of phenomena, and *quantitative*
maps which convey a visual impression of quantity,
value or amount. Each of these main groups can be
sub-divided into maps which show phenomena at a
series of points, those which map features occurring
within specified areas, and those which delineate linear
features such as roads or rivers. These groupings will be
discussed in greater detail in a later chapter on thematic
maps.

THE
EVIL GENIUS OF EUROPE

On a careful examination of this Panorama the Genius will be discovered struggling hard to pull on his Boot. It will be noticed, he has just put his foot in it. Will he be able to wear it?

REFERENCE.

1 Amiens
2 Brussels
3 Berlin
4 Posen
5 Paris
6 Chaum
7 Stutgard
8 Wursburg
9 Narnberg
10 Pilsen
11 Buntzlau
12 Rausbon
13 Innspruck
14 Saltzburg
15 Vienna
16 Maude
17 Aosta
18 Comorn
19 Gap
20 Gratz
21 Montpellier
22 Carcassone
23 Marseilles
24 Toulon
25 Nice
26 Mantua

27 Venice
28 Bologna
29 Spoleto
30 Rome
31 Naples
32 Mostar
33 Callaro
34 Ajaccio
35 The Island

LONDON.   W. CONEY, 61 WARDOUR ST. OXFORD ST.

CARTOGRAPHIC COMMUNICATION

The mapmaking process may be divided into five stages; information gathering; information processing; reproduction; distribution; map use. The first stage, that of *gathering information* (sometimes referred to as data capture or message acquisition) is accomplished in different ways depending on the type of map being produced. Topographic maps involve the acquisition of data by some form of land survey – either by traditional methods of ground survey, by aerial survey or by some form of remote sensing which will incur the use of high flying aircraft or satellites as carriers for the sensing equipment. The acquisition of information for thematic mapmaking follows a different pattern, being based on specialist field survey, the completion of questionnaires, the consultation of census material or other appropriate statistics.

Information processing is probably the most critical of the five stages for much of the success or failure of cartographic communication will depend on the effectiveness with which the processing is carried out. In topographic mapmaking this stage includes the evaluation and editing of information; selection, generalisation and simplification of features; choice and design of symbols; choice of colours, scale, projection, format and so on. Following decisions on these matters, the actual execution of the map (by manual means or by computer) takes place in readiness for the process of map reproduction. Information processing for thematic maps necessitates the evaluation and equating of data. Different countries have different units of measurement; some, for example, provide area statistics in hectares, other in acres. These have to be equated so that the values used for making the final map are strictly comparable. In this respect it is important also that data derived from several sources relate to the same period of time. On many occasions the available statistics are grouped into nominal or numerical classes. In the latter case decisions are made concerning the number of data classes and the points at which the class limits should be set. Several methods of determining the class groupings are available and are fully discussed in a number of books on statistical cartography. Mathematical calculations are often involved; when preparing population density maps, for example, the statistics usually provide absolute population figures within given areas. These figures are converted into ratios of so many persons per unit of area before a density map can be prepared. Computers can greatly assist the mapmaker in processing the large numbers of such calculations. Following his evaluation and manipulation of the data, the mapmaker decides on the most appropriate technique of graphical presentation to use and selects a suitable survey map from which he can derive the basic geographical information necessary to form the framework on which he can construct a thematic map. A camera-ready drawing can then be prepared for the printing process.

The *reproduction* stage can do much to make or mar the visual appearance and clarity of the final map and has a considerable effect on the quality of communication. In the past maps were printed from designs cut in wood blocks or engraved in metal plates; today, topographic maps and atlases are printed by a process called photo-offset-lithography, often in several colours. Thematic maps which usually appear in books and scientific journals are printed either by the letterpress process or by lithography. When letterpress (a relief process) is involved, maps and other line illustrations are reduced photographically to a suitable size for the printed page, the resulting negative being then printed down on to a specially-prepared sheet of metal which forms the printing surface. This metal plate is etched in acid so that the design to be printed stands out in relief. It is then mounted on to a wooden block at type height ready for printing along with the text of the appropriate book or journal.

The process of photo-lithography by which a great many illustrations for books are reproduced today (including the present volume), begins in the same way as that for a line block, with a reduced negative prepared from the original drawing. The negative is printed down on to a litho plate (which may be metal, plastic or even paper), the image being fixed chemically so that when printing ink is rolled over the plate it adheres only to the image lines. During printing the plate is attached to a rotating plate cylinder and water applied to it by dampening rollers, making the plate grease-repellent except for the image lines which reject the water. Inking rollers apply greasy ink to the plate and this adheres to the image but is rejected by the moist non-printing areas. Printing paper is then pressed against the plate by an impression cylinder and receives the inked image.

Stage four, the *distribution* stage, is in a sense outside the cartographic process though it is part of the communication cycle. It involves the checking of proofs, editing, market research, marketing and distribution by a publisher to retail outlets. Stage five, however, that of

Fig 8 **A 19th-century political cartoon map,** *The Evil Genius of Europe* **(1859), the character in question being Napoleon III who had signed a treaty with Piedmont in order to bring about a war with Austria and remove her from the parts of Italy she occupied.**
Courtesy: **British Library Board   BL 1078 (24)**

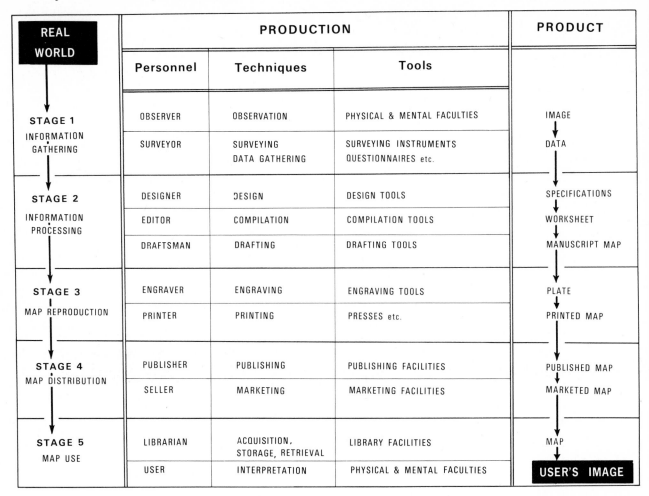

REAL WORLD	PRODUCTION			PRODUCT
	Personnel	**Techniques**	**Tools**	
STAGE 1 INFORMATION GATHERING	OBSERVER	OBSERVATION	PHYSICAL & MENTAL FACULTIES	IMAGE ↓ DATA
	SURVEYOR	SURVEYING DATA GATHERING	SURVEYING INSTRUMENTS QUESTIONNAIRES etc.	
STAGE 2 INFORMATION PROCESSING	DESIGNER	DESIGN	DESIGN TOOLS	SPECIFICATIONS ↓ WORKSHEET ↓ MANUSCRIPT MAP
	EDITOR	COMPILATION	COMPILATION TOOLS	
	DRAFTSMAN	DRAFTING	DRAFTING TOOLS	
STAGE 3 MAP REPRODUCTION	ENGRAVER	ENGRAVING	ENGRAVING TOOLS	PLATE ↓ PRINTED MAP
	PRINTER	PRINTING	PRESSES etc.	
STAGE 4 MAP DISTRIBUTION	PUBLISHER	PUBLISHING	PUBLISHING FACILITIES	PUBLISHED MAP ↓ MARKETED MAP
	SELLER	MARKETING	MARKETING FACILITIES	
STAGE 5 MAP USE	LIBRARIAN	ACQUISITION, STORAGE, RETRIEVAL	LIBRARY FACILITIES	MAP ↓
	USER	INTERPRETATION	PHYSICAL & MENTAL FACULTIES	USER'S IMAGE

Fig 9 **Diagram showing the stages of the mapmaking communication process (after Woodward).**

map use, is an integral part of the communication process. It concerns the acquisition, storage and retrieval of maps, either by a map curator acting for a large library, or by individual users. The final step in communication is for the map to be observed, read and interpreted by the map user. If communication has been accomplished successfully, the picture received by the user will correspond with the view of reality the mapmaker wished to convey.

The entire process reviewed briefly above is one of graphic communication, a concept on which the attention of cartographers is increasingly being focussed, in contrast with past concentration on the technical and practical sides of mapmaking. Only now, it is suggested, is sufficient attention being paid to the uses of maps and the needs of the map user. This has led not only to market research into users' preferences and requirements, but also to research in less obvious directions

such as the development of tactile maps to aid the mobility of blind people.

In the earliest times, certainly prior to the application of printing to the mapmaking process, cartography was a straightforward person-to-person communication with a map drawn in manuscript by a scholar, traveller or mariner to convey his own ideas or those theological concepts he wished to illustrate. The introduction of printing, the development of relatively sophisticated survey techniques and the evolution of a complex system of symbolisation complicated the pattern, and today the association of advanced technology at each stage of the cartographic process, together with the infinite variety of topics to be mapped, has produced a more complex visual communication system.

Numerous attempts have been made to provide a rational analysis of cartographic communication and resort has been made to analogies with other disciplines, notably electronics. A loose analogy may be made to electronics communication theory in which a signal is transmitted to a receiver via a communications channel. If this is translated into mapmaking terms, we have a

to half the Equator in length and spaced at uniform angular intervals. Thirdly, each meridian intersects each parallel at 90°. A map projection is merely a means of representing these parallels and meridians on a flat map but no one projection preserves all the properties mentioned above and some form of distortion is unavoidable.

The most important points to be considered are: first, the location of a point or area in relation to the whole of the earth's surface; secondly, the direction which any point bears to another; thirdly, distance between points; fourthly, the accurate representation of areas. Few people are called upon actually to construct map projections these days, and the main interest of the mapmaker is in the relative advantages and disadvantages of different projections in communicating geographical information. One type of projection may reproduce the shape of land masses accurately (a *conformal* projection); another reproduces relative areas accurately (an *equivalent* or *equal-area* projection); yet another maintains true direction (an *azimuthal* projection). The mapmaker's task is to select that projection which possesses the positive qualities he requires for the work in hand. If his concern is with navigation or some form of migration he will look for one which preserves correct direction or accurate distances, but if he wishes to present a realistic view of a geographical distribution over large areas he will need an equal-area projection in order to convey an accurate picture of density throughout his map.

Projections may be generally classified according to the way in which they are produced. The first group are the strictly geometrical projections which are derived from a 'generating' globe by actual processes of projection. These are termed *perspective projections* and they can be varied in type by altering the position of the point of origin of the projection – making it the centre of the generating globe, a point on its surface or a point outside the globe at an infinite distance. Variety is also achieved by varying the nature of the surface on which projection is made; it can be a *plane* surface usually tangential to the globe at a specified point; a *cylinder* which can be placed around the globe or intersect it in a particular manner; a *cone* which can rest upon the globe or intersect it. The second group of projections are derived from their perspective counterparts by some modification and are known as *non-perspective* projections. The degree of modification can be arranged to meet with a particular requirement so that it is possible to ensure that areas are strictly comparable whether the projection is conical, cylindrical or zenithal. Alternatively modifications can ensure that the scale at any point is the same *in all directions*, although the actual scale varies from one part of the projection to another. This type of projection is termed *orthomorphic* and preserves the shape of small areas. A third group of projections consists of those in which the parallels and meridians are drawn so as to conform to an arbitrarily selected principle. The concept of 'projection' is not readily apparent and the projections do not consist of modifications of any other projection. This group includes a number of projections which display the whole world on one sheet – the elliptical *Mollweide*, for instance, which is an equal-area projection often used in small atlases.

Plane projections

When the network of parallels and meridians is geometrically projected on to a plane which is tangential to the globe and at 90° to the line joining the point of origin to the point of contact several projections result, depending on the position of the point of origin. They have one common property – direction, or *bearing*, from the map's centre is true and they are therefore *azimuthal* or *zenithal* projections. The mode of projection can be easily understood by imagining a glass sphere on which the graticule is clearly marked and a light source placed at the appropriate point of origin. When the light is positioned at the centre of the globe and a sheet of paper is made to touch a pole, tangential to the sphere, a shadow of the graticule will be cast on to the paper with the meridians projected as straight lines and the parallels as concentric circles, increasingly widely spaced with distance from the pole. The Equator itself is not projected as it is at 90° to the light source but can be made to appear by moving the light from the centre of the sphere to the remote pole. The first of these versions of the azimuthal group is termed *Gnomonic*; when the plane of projection is tangential at a pole the resulting projection is referred to as a *polar* case; when it is tangential at a point on the equator the projection is an *equatorial* case; when tangential at any other point it is an *oblique* case. All cases possess an important property in that great circles are projected as straight lines, a valuable property in the mapping of aeronautical routes. The second version, in which the point of origin is on the globe's surface opposite the point of contact, is termed *Stereographic* and once again, polar, equatorial and oblique cases are possible. This is a more useful version for the mapping of larger areas as distortion away from the centre is less than with the Gnomonic. It is popular, therefore, for polar maps, hemispheres and continental maps. A third version of the azimuthal group can be obtained by moving the viewpoint to an infinite distance. This is known as an *Orthographic* projection and the plane of projection can be tangential at any point so that polar, equatorial

CYLINDRICAL

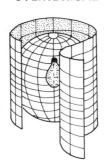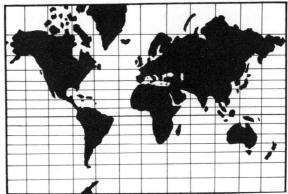

When a light source is placed at the centre of a translucent globe and a cylinder of paper
wrapped around the globe, the parallels and meridians are projected
as straight lines intersecting at 90° to produce a cylindrical projection

AZIMUTHAL

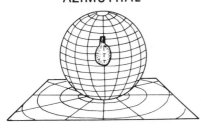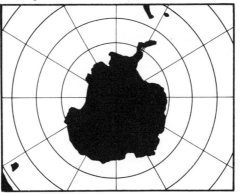

When the meridians and parallels are projected on to a plane surface tangential to the globe
at any point, often a pole, an azimuthal or zenithal projection is produced in which direction
is always true from the central point

CONICAL

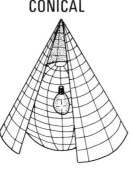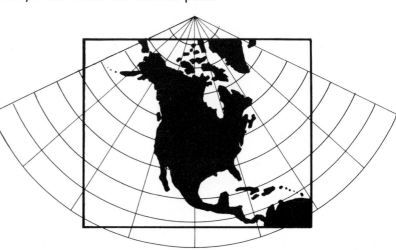

If a paper cone is placed over the globe, touching it a parallel,
the meridians are projected as straight lines
and the parallels as arcs of concentric circles

Fig 17 **Three types of perspective projection.**

and oblique cases are practicable. The resulting projection resembles a very distant view of the globe as it might be seen from an orbiting satellite.

Cylindrical projections

In the most common form of cylindrical projection the cylinder is placed around the globe, cut at a suitable point parallel to the axis and opened flat. The cylinder's axis normally coincides with that of the globe and the cylinder can touch the globe along the equator or intersect the globe along two parallels of latitude. Normal cylindrical projections have certain common characteristics; the parallels are projected as parallel straight lines of equal length, one of which is correctly divided for the intersections with the meridians; meridians are projected as parallel straight lines, of equal length and equally spaced; all meridians and parallels intersect at 90°; the projections are rectangular in shape. There is considerable scale distortion along those parallels which are far distant from the standard parallel but the scale along the meridians can be made correct as all are equal in length, just as on the globe.

The most familiar cylindrical projection, the *Mercator*, is a non-perspective or modified projection. Its principle is that parallels of latitude are made equal in length to the Equator so that the scale along them becomes more and more exaggerated with distance from the Equator; the distances of the parallels from the Equator are then adjusted so as to make the scale along the meridians at any point equal to the scale along the parallels at the same point. In other words the stretching of the world in an east-west direction is accompanied by an equal amount of north-south stretching, the amount of stretch varying with latitude. This means that at every point there is true representation of shape but in the case of large areas shape, as well as scale, is highly distorted. It can be enlightening, for example, to compare the representations of Greenland and South America on a Mercator map. The projection makes Greenland appear to be the larger, whereas in fact it is nine times smaller than South America. Because of the widely varying scales at different latitudes the reader is normally provided with a composite scale which has bars showing the true scale at the Equator plus the map scale at several latitudes. There is little doubt that many people's world view is the misleading one revealed by Mercator. This is unfortunate, of course, since the declared use of the projection was for navigation, its most important property being that a straight line on the projection is a line of constant

bearing. Mercator himself wrote in connection with his great world map of 1569 that

> If you wish to sail from one part to another here is a chart, and a straight line on it, and if you follow this line carefully you will certainly arrive at your destination. But the length of the line may not be correct. You may get there sooner or may not get there as soon as you expected, but you will certainly get there.

In the standard Mercator projection the axis is its contact line with the globe i.e. the Equator. On either side of that axis for 15° north and south the Mercator provides a reasonably accurate map. Alternatively the cylinder can be turned half way round and a selected meridian used as the point of contact. Once again a narrow band of fairly accurate portrayal is provided, this time between 15° east and west of the chosen meridian. The advantage of the *Transverse Mercator*, as this modified version is called, is that meridians can be selected which cross greater extents of land and more densely populated areas than those crossed by the Equator. The properties of the projection can thence be utilised to greater advantage.

Conical projections

These are visualised as being produced on a conical surface which is afterwards opened out flat. In the usual form the cone's axis is coincident with that of the globe over which it is placed. The apex of the cone is over a pole and the cone itself touches the globe along a parallel, usually between 40° and 60° north or south of the Equator. Meridians are projected as straight lines and parallels as arcs of concentric circles. Distances are correct along the parallel touching the cone, termed the *Standard Parallel*. Away from it the scale becomes increasingly distorted along both meridians and parallels.

The *Simple Conic* is a non-perspective version based on one standard parallel and is only suitable for a map of a narrow belt along this parallel. The *Two-standard* Conical Projection is a modification in which a cone intersects the globe along two selected parallels of latitude. The scale along these parallels is true, for the area between them it is too small, outside them it is too large. This version does, however, provide a greater area in which the scale is reasonably accurate than when only one parallel is used. By a judicious choice of parallels it is possible to achieve a reasonably good map of a large area, providing the latitudinal extent is not great.

Several other modifications of the conical projection are available including the *Bonne* which is familiar to users of European atlases because it gives a fair representation of shape for countries in middle latitudes as

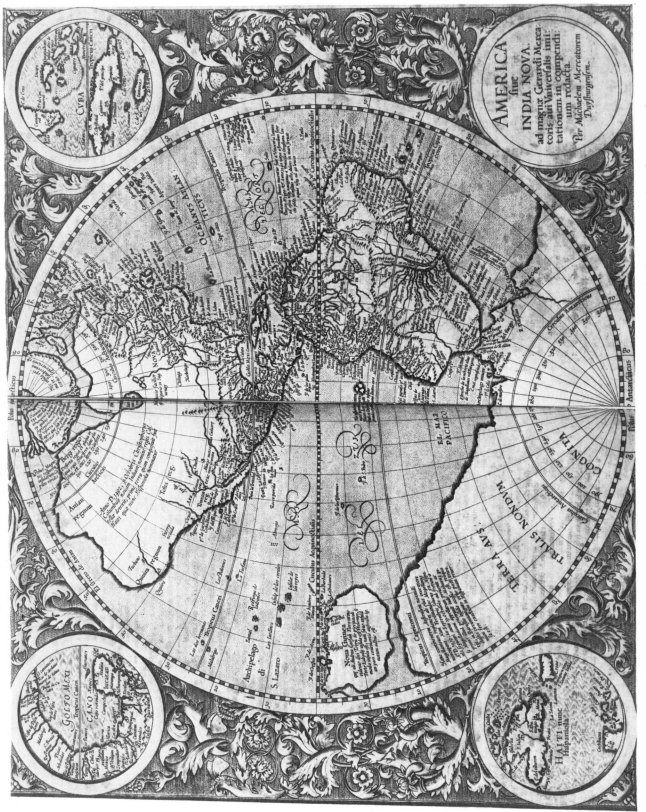

Fig 18 Michael Mercator's map of the Americas (1606) prepared on a stereographic projection. The map includes enlargements (no scale provided) of the Gulf of Mexico, Cuba and Haiti. It is also noteworthy for the elegant ornamentation, reminiscent of silversmiths' work, between the map and its frame.
Courtesy: University of Liverpool   Ryl.N.1.9

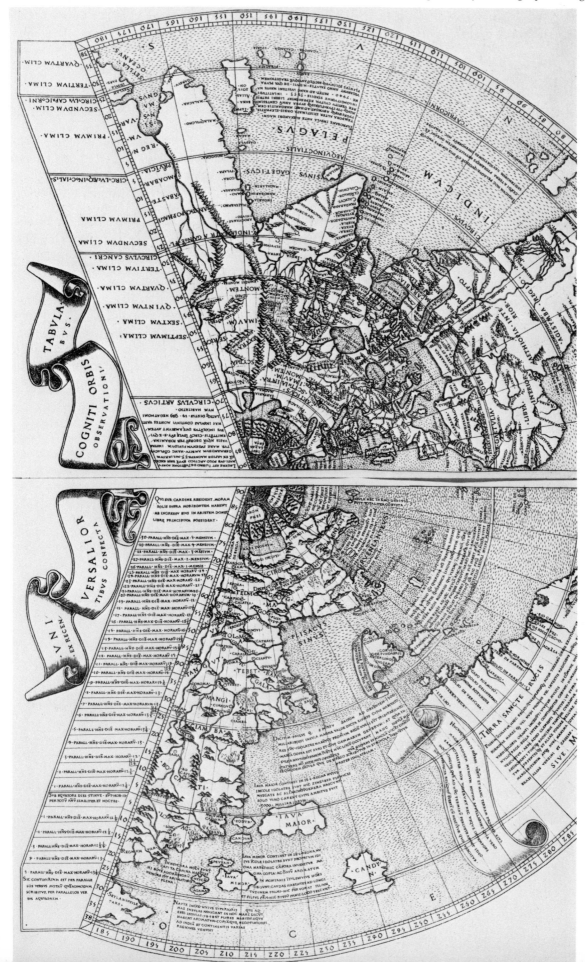

Fig 19 Johan Ruysch used a conical projection for his world map in Ptolemy's *Geographia* (Rome, 1508). The conical projection was introduced by Ptolemy but its use is generally confined to the portrayal of areas of limited latitudinal extent. Ruysch's attempt to present both northern and southern hemispheres on the single cone have resulted in considerable distortion in southern latitudes. From a facsimile in A. E. Nordenskiöld's *Facsimile-Atlas*. *Courtesy:* University of Liverpool 202.1.2

well as showing strictly proportional areas. It is more effective for areas of greater north–south than east–west extent.

Interrupted projections

'Interrupting' a projection is a technique sometimes employed to achieve better representation of shape and smaller scale error and particularly to display the land masses of the earth to greater advantage. Interruption usually takes place, therefore, in an area of sea. The only proviso necessary for a projection to be interrupted and re-centred is that the linear scale along each parallel be uniform and therefore that the parallels be uniformly sub-divided by the meridians. A number of central meridians are chosen and the projection grid is constructed around each. The method can perhaps be better understood by referring to a globe. It will be noticed that the surface of the globe is achieved by pasting paper on which the map is printed to the sphere. This spherical surface is made up of a series of *gores*, each of which has a central meridian running through it in a north–south direction and the Equator at its midpoint which serves to hold all the gores together. One of the most successful interrupted designs is the *Goode's Homolosine Equal-Area Projection* which can be arranged to give an equal-area map of the world's ocean basins or to give a good representation of continental masses.

Within the space available it is possible only to scratch the surface of such a broad topic as map projections. There is, however, a considerable volume of specialised literature to which the reader who wishes to investigate further is directed. Despite the great number of available projections new ones are still being devised. One of the most recent is the *Peters Projection* designed by the German historian, Arno Peters. It is an equal-area projection which has been used to good effect in a poster publicising the work of Christian Aid. The aim of the map is to depart from Mercator's world view in which the earth is distorted in favour of the countries occupied by white people. The Peters Projection places the Third World countries firmly and prominently at the centre of the map and draws attention away from Europe. In appearance, however, the Peters world view is decidedly odd due to the land surfaces close to the Equator being elongated in a north–south direction whereas those in high latitudes are compressed – Africa, for instance, is portrayed as an elongated, narrow continent. Nevertheless the new map provides a good example of the development and use of a completely new projection to convey a specific concept.

3 FUNDAMENTALS OF MAPMAKING: (ii) The development of a cartographic vocabulary

Any map as an entirety is a symbol of the real world or some portion of it. It also comprises a great variety of smaller items which themselves symbolise some aspect of the environment. Environmental phenomena occur at a series of points, along specific lines or within certain areas and three basic classes of symbol are required to cope with these situations. In addition to these 'class' symbols, each of which has a precise, fixed location, most maps include names. By adding a name alongside a particular symbol it is transformed from a class symbol into a unique feature. A small circle might be used to indicate the position of a town, for example, but when the name 'Oxford' is placed in close relation to it, the circle acquires a unique character with the name and circle forming a composite symbol which indicates that Oxford is situated at this precise location and cannot occur elsewhere.

One of the first things to understand about cartographic symbols is that they can not only tell us the nature of a particular feature – what it is – but they can also be made to indicate quantity, value, size, height, intensity and many other factors. The first type of symbol, in which no element of quantity or value is indicated, is termed 'qualitative'; the second type is referred to as 'quantitative'. Symbols can also be *arbitrary*, with their appearance bearing little or no relation to the feature being represented or they can be *pictorial* with the symbol suggesting the natural appearance of the relevant feature at a reduced scale. Despite the complexity of the contemporary alphabet of cartographic symbols – and unlike the verbal alphabet the possible number of characters is limitless – particular symbols generally refer only to a general group of phenomena and are unable to make sophisticated distinctions within that *genre*. This can be illustrated by referring to the familiar symbol of a square surmounted by a cross which symbolises a church with a tower quite successfully. If it became necessary to extend the meaning of this symbol to represent other features of the church difficulties would immediately arise. How for instance could the

period of building be indicated, or the building material, or the seating capacity? It has to be accepted that in any system of map symbolisation there must be limitations in the amount of information it is possible to communicate about a particular item.

Needham remarks that many geographical symbols of great antiquity are embodied in the Chinese language (*Science and civilisation in China*, vol. 3 (Cambridge, 1959), p. 498). The character for a river is an ancient graph of flowing water; that for a mountain was once an actual drawing of a mountain with three peaks; that for fields illustrates enclosed and divided spaces; political boundaries appear in the character for country. Bone and bronze forms of the character for 'map' really show a map and Needham suggests that the very pictorial character of the Chinese language did much to encourage the idea of mapping. Certainly it is true that early Chinese maps were essentially pictorial with some of them showing much of the delicacy of execution seen in Chinese painting.

The content of a topographical map comprises natural phenomena such as mountains, rivers and lakes along with man-made items such as towns, roads, railways, factories, windmills, public buildings. An extensive system of conventional symbols has been developed over many centuries in order to convey graphically the location and identity of such features.

HILLS AND MOUNTAINS

The mapping of the three-dimensional character of the earth's surface has always posed problems. Man's earliest known attempt to depict a range of mountains was in the Babylonian clay tablet map of Northern Mesopotamia (*c*. 3800 BC) on which two flanking mountain ranges are suggested by fish-scale-like symbols. An interesting point about the general symbolisation on this map is that the compiler has tried to show the mountains in profile but has represented all other features as they would be seen from above in plan view.

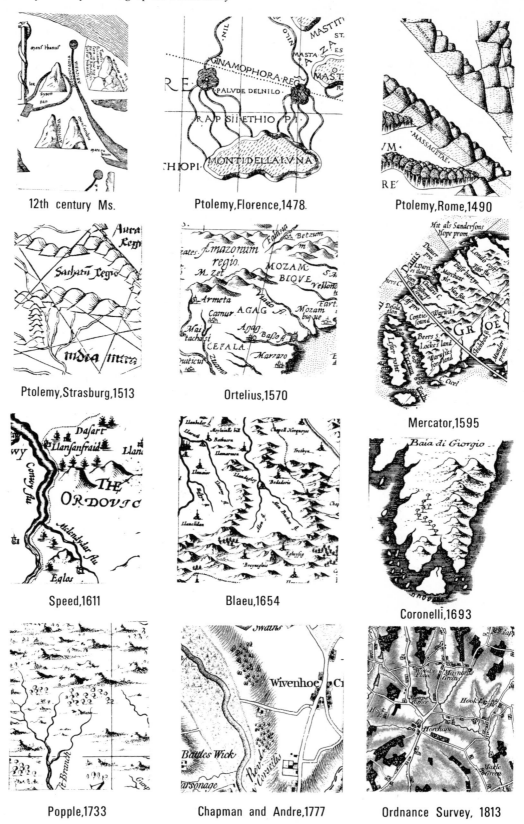

12th century Ms.

Ptolemy,Florence,1478.

Ptolemy,Rome,1490

Ptolemy,Strasburg,1513

Ortelius,1570

Mercator,1595

Speed,1611

Blaeu,1654

Coronelli,1693

Popple,1733

Chapman and Andre,1777

Ordnance Survey, 1813

Fig 20 **Some ways in which mapmakers have attempted to provide a graphic impression of relief prior to the introduction of contours on to land maps in the 19th century.**

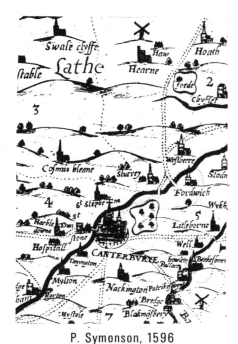

P. Symonson, 1596

William Smith, 1602

J.B. Homann, 1716

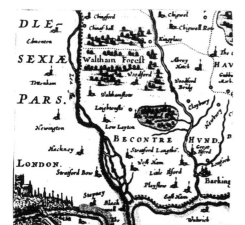

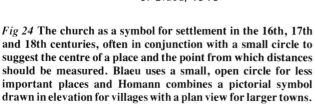

J. Blaeu, 1645

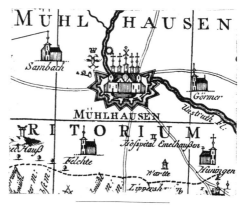

P. Schenk, 1760

Fig 24 **The church as a symbol for settlement in the 16th, 17th and 18th centuries, often in conjunction with a small circle to suggest the centre of a place and the point from which distances should be measured. Blaeu uses a small, open circle for less important places and Homann combines a pictorial symbol drawn in elevation for villages with a plan view for larger towns.**

in which the message is conveyed. On early woodcut maps the craftsman used a small open circle to symbolise a settlement. This was not only comparatively easy for him to execute but the centre of the circle provided a precise locational point from which distances between towns could be measured. By the mid sixteenth century

German makers of woodcut maps had developed a number of symbols for different types of settlement. This meant that maps had to include a table which explained the meaning of each symbol to the map user. Such a table is usually referred to as a *key*, *legend* or *reference table*. Philip Apian, a German cartographer, is famous for his large-scale woodcut map of Bavaria (1568), a map which displayed a comparatively sophisticated treatment of settlements; pictorial symbols represented free towns, bishoprics, monasteries and villages while towns, markets and castles had arbitrary symbols based on varying treatments of a circle. Apian also used

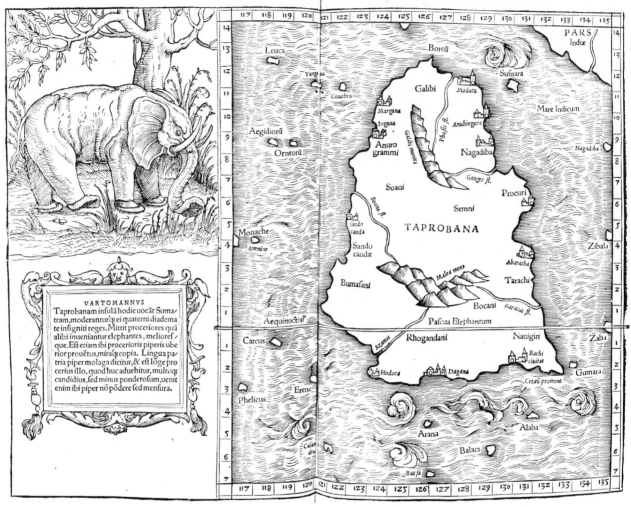

TABVLA ASIAE XII·

VARTOMANNVS

Taprobanam infulã hodie uocãt Suma-
tram, moderanturᵭ ei quaterni diadema
te infigniti reges. Mittit proceriores quã
alibi inueniantur elephantes, melioref-
que. Eſt etiam ibi procerioris piperis ube
rior prouĕtus, miraᵭ copia. Lingua pa-
tria piper molaga dicitur, & eſt lõge pro
cerius illo, quod huc aduehitur, multoᵭ
candidius, ſed minus ponderoſum, uenit
enim ibi piper nõ põdere ſed menſura.

Fig 25 **Sebastian Münster supplemented the somewhat bare map of Taprobana in his edition of Ptolemy's *Geographia* (Basle, 1540) with a fierce looking elephant and indicated elephant pastures (*Pascua elephantum*) on the map. Considerable confusion existed in the 16th century as to whether Ceylon or Sumatra was the island of Taprobana seen on Ptolemy maps. In this case Sumatra was indicated in the panel of descriptive notes by the traveller, Locovico di Varthema or Vartomannus.** *Courtesy*: **University of Liverpool Ryl.N.2.14**

pictorial symbols for industrial locations such as glass-works and saltpans. All these items were explained in the clearly-arranged legend.

In the late sixteenth and early seventeenth centuries churches were adopted as the standard symbols for a village, a natural choice as the church with its prominent tower or spire was generally the building which stood at the core of the settlement. Philip Symonson's remarkable county map of Kent (1596), one of the finest county maps of its period, shows an attempt by Symonson to portray each church faithfully, his tiny engravings show-ing, in elevation, churches with spires, those with towers and so on. He also featured picturesque symbols for castles and windmills and showed great houses with their parks surrounded by a pale fence. Large towns such as Rochester and Maidstone were effectively depicted by tightly-packed clusters of houses.

During the seventeenth century small bird's eye views were often used to indicate settlements, but from the eighteenth century it became more usual to depict them in plan. The open circle, however, remained in vogue to represent hamlets and for settlements on very small-scale maps. The tower or building symbol did not entirely disappear as yet and even in the closing decades of the eighteenth century maps are to be found in which this technique was perpetuated. Today it is customary to show towns in plan on topographical maps, the extent of the built-up area, pattern of major streets and prominent buildings being indicated. Atlas maps, on the other

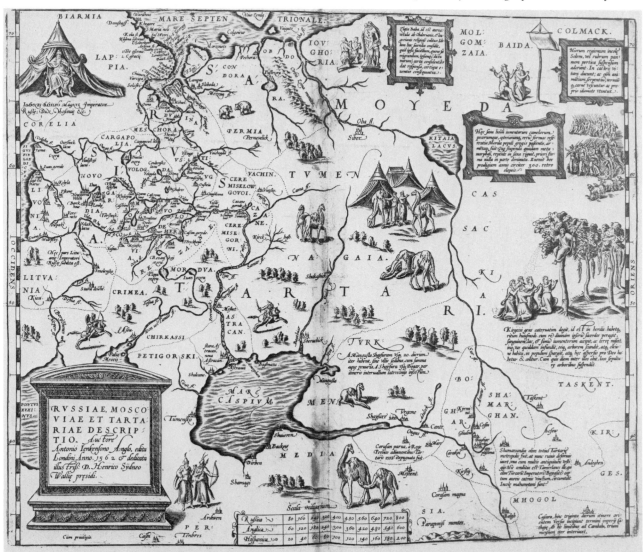

Fig 26 Anthony Jenkinson led an expedition to Russia in 1557 with the object of opening up trade routes for the Muscovy Company. A map which he completed was used in the *Theatrum Orbis Terrarum* (1570) of Ortelius. The map is especially noteable for its pictorial indication of fauna and for the way in which Jenkinson used verbal descriptive notes to supplement his cartographic message. The scale of miles is unusual and Jenkinson indicates degrees of latitude but not longitude.
Courtesy: University of Liverpool   Ryl.N.1.3

hand, often feature geometrical symbols, usually variants on a circle or square, the treatment indicating the rank of the town, e.g. a small circle for a minor place; a circle around a dot for a bigger town; a large filled-in circle for a city with a considerable population and so on. Sometimes a more precise quantitative approach is needed to communicate information about urban populations and on some maps symbols of graded size are used to represent numbers of people. Alternatively, the areas of the symbols can be made proportional to the population represented.

WOODLAND, FLORA AND FAUNA

On early manuscript maps it was customary to represent woodland and forest by clumps of trees drawn in elevation, little attempt being made to indicate the shape or extent of the area covered. The maker of the Bodleian Map (*c.*1360) symbolised Sherwood Forest and the New Forest by a pair of intertwined trees and the pictorial symbolisation was carried over into the era of engraved maps. Particularly interesting treatments of woodland can be seen on the maps in the numerous printed editions of Ptolemy's *Geographia*. In the late fifteenth and early sixteenth centuries there were attempts at distinguishing coniferous and deciduous woodland – Münster, for example, symbolised the different types in

Ptolemy, Rome, 1490

Ptolemy, Strasburg, 1513

Mercator, 1538

Munster, 1545

Ortelius, 1570

Blaeu, 1645

De Lisle, 1768

Schenk, 1760

Ordnance Survey, 1824

Fig 27 **Some examples of calligraphic styles used on maps from the 15th to the 19th century.**

a manner which is instantly recognisable, and Olaus Magnus in his splendid pictorial map of Scandinavia (1539) included handsome coniferous trees among his miniature pictures showing the costumes, customs and fauna of the northern lands. As time went on there was a gradual trend towards engraving smaller, more delicate trees with shadows added to the east, a technique well seen in the maps of Blaeu, *c.*1635.

Early maps of European countries often included a conventional symbol to locate vineyards and the chosen symbol remained in remarkably consistent use for over two centuries from its introduction by Münster in the mid sixteenth century. The symbol devised was a simple one which stylistically illustrated a vine snaking upwards around a vertical pole.

British Ordnance Survey maps have generally made a distinction between deciduous and coniferous woodland and it seems a retrograde step to eliminate this differentiation from the current 1:50,000 series. At the moment this series distinguishes wood, orchard and park or ornamental grounds in contrast to an Istituto Geografico Militare 1:50,000 Italian sheet which shows the boundaries of woodland and distinguishes firs, pines, cypresses, eucalyptus, cork-oaks, oaks, elms, chestnuts, beeches, larches, poplars, vineyards, orchards, citrus groves, olive groves, almond trees, scrub, reafforestation and coppice. A further distinction is made between open wood (shown by one symbol) and thick wood (by three grouped symbols). Such a remarkable variety of information gives some indication of the degree of sophistication in symbolisation which can be achieved on modern maps.

The indication of animal life was an attractive feature of early mapmaking, for explorers, encountering species never met with in Europe, naturally wished to record them cartographically. Some of this portrayal of fauna has been derided as mere space filling – perhaps justifiably in regard to sea areas – but Wilma George suggests that there is evidence to show that many of the species and their habitats were mapped with some accuracy and their indication thus provides useful historical evidence for the modern biologist. Generally speaking it was the larger, more exotic animals which attracted mapmakers, and lions, elephants, crocodiles and colourful birds abound on sixteenth- and seventeenth-century maps.

Fig 28 Münster's map of the New World from his edition of Ptolemy's *Geographia* (Basle, 1540). As in many other woodcut atlases some of the lettering was executed by inserting metal type into the wood block. Münster is careful to warn travellers of cannibals in Brazil and he locates a race of giants in Patagonia.
Courtesy: University of Liverpool   Ryl.N.2.14

Fig 29 **Elaborate title cartouche from John Mitchell's** *A Map of the British Colonies in North America* **(1755) with an attractive blend of typographic styles.**
Courtesy: **Harry Margary and the University of Cambridge**

LETTERING

The addition of geographical names as a means of identifying features is an important part of map design and the quality of the lettering used for this purpose can play a major role in determining a map's visual effectiveness. Lettering on a map serves other purposes than mere identification; by the way in which it is spaced and arranged it can provide some indication of the linear extent of mountain ranges or the areal extent of national territories or administrative areas; by varying the size, character or colour of the lettering the class to which the named feature belongs can be implied. Rivers, lakes and seas, for example, are normally lettered in blue while the lettering used to name settlements is varied in style and size to indicate population or status.

In the period of manuscript maps all lettering had to be executed by hand, and styles varied enormously from contemporary book-hands such as the Charlemagne or Gothic hand used on the Hereford Map to the beautiful lettering based on the old Roman inscription lettering rediscovered by Italian Humanists in the 1470s and exquisitely executed by Diego Ribeiro on his world chart of 1529. Roman capitals were the only type of lettering used on some of the Ptolemy maps prepared between 1477 and 1490 but were later used in conjunction with other scripts. In general the makers of early woodcut maps favoured the Germanic Gothic or Black-letter scripts but small lettering was extremely difficult and an ingenious technique devised to overcome the lettering problem was to set names in the recently introduced movable metal type, cut holes in the wood block, and wedge the metal type into the holes so that it would print with the other map detail.

With the introduction of engraving on copper sheets the problem of lettering was reduced although the craftsman was faced with the task of cutting letters in reverse in the copper plate. Mercator devised his own *italic* hand especially for mapwork and used this smoothly-flowing style alongside well-proportioned Roman lettering. Blaeu and Jansson in the great period of Dutch cartography added a new elegance to lettering and a feature of Dutch seventeenth-century maps is the flourishing 'swash' lettering with which mapmakers liked to fill in unwanted areas of space. The calligraphy of Jodocus Hondius as seen on the British county maps of John Speed and William Smith was notably well formed and executed, fulfilling admirably the cartographic requirements of legibility, perceptibility and suitability.

The italic hand remained popular during the eighteenth century but other lettering, in which the engravers attempted to ape the type designer, was undistinguished. During the eighteenth and nineteenth centuries map titles became a vehicle for the display of an extraordinary variety of styles, well seen in the large-scale county maps by the Greenwoods and in John Mitchell's *Map of the British and French Dominions in North America* (1755). Lettering used on the face of maps, however, became plainer with mapmakers striving for clarity. The lettering of the charts in J. F. W. Des Barres' *Atlantic Neptune* (1775–81) is noteworthy for the way in which the engraver has varied the styles and sizes of his lettering to indicate both location and relative importance without any sacrifice in aesthetic quality.

The complexity of modern maps necessitates the devotion of much thought to the choice of types used. The Ordnance Survey 1:1250 and 1:2500 maps produced before 1879, for example, included no less than nineteen different styles of lettering simply to indicate administrative areas and settlements. The Seventh Series 1:63,360 maps (recently superseded by the 1:50,000) indicated towns of different sizes by five sizes of the Times New Roman type face and the smallest towns (under 10,000) by Times Italic. Thirty-four variations of type style or size were used to name features ranging from parishes, villages, hamlets, farms, to lakes, rivers, forests and sandbanks. It can be a very enlightening exercise to compare national map series of different countries, noting the quality of the lettering and assessing its effect on communication as a result of well-chosen sizes and weights of type, pleasing type designs and so on.

This brief survey of the evolution of only some of the commoner types of symbol may have suggested by implication that there is virtually no limit to the phenomena that may be symbolised on a map. It is of great importance from the map user's point of view that all symbols be well designed, be as evocative as possible of the feature they represent, and be readily distinguishable from other symbols.

4 MAKING THE MAP

ACQUIRING THE MESSAGE AND PROVIDING THE DATA

Before a topographical map can be produced the area under consideration must be subjected to detailed survey. Surveying is concerned with making detailed measurements from which an accurate representation of the earth's surface features can be made, to scale, on a map. For surveys of a small area the errors introduced by assuming the earth's surface to be flat rather than spherical are negligible and it is only necessary to determine the relative positions of points. Such a survey is termed a *plane survey*. For larger areas, e.g. surveys of a national character, the earth's curvature must be taken into account and the absolute locations (latitude and longitude) of control or key points determined by astronomical observations and by triangulation. This branch of surveying is referred to as a *geodetic* survey.

Surveying, therefore, which provides the data for a very accurate graphic representation of part of the earth's surface, is the first stage in the process of communicating geographical information visually in map form. It supplies the framework on which the data appropriate to a particular purpose can be presented.

Surveying in the ancient world

Few records are available about the instruments or methods of survey used in ancient times but existing evidence suggests that some form of land surveying took place many centuries ago in the valleys of the Nile, Tigris and Euphrates which have long been regarded as the birthplace of western cartography. Plans of cities and surrounding agricultural areas are recorded on the clay tablets of early Babylonia, and boundary stones which demarcated the boundaries of fields and pasture still exist after some 3000 years. The walls of tombs of prominent Egyptian citizens bear records of land surveys, the inscriptions providing information about the dimensions of plots, quality of land and taxes payable on each prop-

erty. Some insight is given into the kind of instruments used in making the surveys. Cords and rods were employed for linear measurements, the cords being knotted at intervals to mark off the units of length. The Egyptians also had a device called a *groma* which was used in laying out right angles, both for the division of land into rectangular plots and in the setting out of road networks. In China similar instruments were in use and literary references speak of compasses and squares, plumb-lines, ropes, cords and graduated poles.

It is to the Greeks and Hero of Alexander, in particular, that we owe the first systematic approach to classical surveying. Hero prepared a treatise which set down the basic principles of surveying and described the *dioptra*, an instrument which had applications as a level or for setting out right angles. The Romans used wooden rods and knotted cords for linear measurement in addition to folding rules similar to those in use today. They improved the instruments of the Greeks and developed their own along with surveying techniques which were needed in administrative fields such as land division and road construction.

As used today the term *levelling* relates to the determination of relative heights of points on the earth's surface. Some form of levelling must have been used in Roman times for laying out gradients of drainage channels and two kinds of levelling instruments are recorded. The first, called *chorobates*, consisted of a trough partially filled with water and depended on the principle that when at rest the surface of a liquid is horizontal. The second incorporated a plumb bob and relied on the principle that lines perpendicular to a vertical line are horizontal.

The Middle Ages

One of the most important surveying instruments of the Middle Ages was the *cross-staff*, an instrument thought to have been developed in China, which was a graduated rod of rectangular cross-section with a cross-piece which

smaller-scale final map and the cartographer has then to use his experience and skill in simplifying coastlines, boundaries and other features while at the same time preserving their major characteristics. If maps of smaller scale are enlarged to construct the final map there will be an undesirable reduction in accuracy as the base maps will already have been generalised to a greater degree than is required.

THE MEDIUM OF CARTOGRAPHY – PREPARING AND REPRODUCING THE MAP

The baked clay tablets on which were recorded the cuneiform writings, maps and plans of scholars in Mesopotamia thousands of years ago offered the advantage of durability – witness the number of surviving examples – but were ineffective as a means of disseminating information due to their lack of portability. Their use was restricted to a minority group of scholars, priests and administrators. Papyrus, grown on the Nile marshes, provided a more practicable writing material from the point of view of spatial transmission but lacked the necessary durability for mapmaking. Furthermore, during the period of Islamic expansion in the eighth century, exports of papyrus were cut off and parchment became the main material used in European book production. This material, usually made from the skin of a sheep or goat, and vellum, a fine parchment originating from calf skin, offered an acceptable surface for drawing, colouring and illumination and was widely used in medieval cartography. The standard portolan chart, for example, was prepared on a skin of parchment and the large medieval *mappae mundi* of Hereford and Ebstorf were also prepared on this medium.

Parchment and vellum were excellent for writing and the making of maps in manuscript but were unsuitable for the printing process on its introduction into Europe during the fifteenth century. The innovation of printing maps from woodblocks and engraved copper sheets is regarded as the watershed of cartographic history but the equally significant role played by the introduction of papermaking into Europe is often overlooked. A prerequisite of printing and the distribution of printed material was an economical, light, tough, readily-available material with a surface which would accept ink without penetration and would stand up to the rigours of the printing press. Paper had been invented in China in AD 105 by Ts'ai Lun who used old rags, fishing nets and rice stalks as his raw materials. These were ground to a pulp, washed, beaten and mixed in a vat to which sizing matter was added to give the paper a non-absorbent surface which would take ink. The mixture was agitated for some time before the paper sheet was produced on a flat

mould made from a mesh of bamboo strips. Paper was used in China for block printing in the eighth century and for book printing a century later. The technique of paper manufacture spread slowly westward via Samarkand and Baghdad in the eighth century, thence to Egypt, and was introduced into Europe via Spain in the tenth century, into France in the twelfth, Italy in the thirteenth, Germany in the fourteenth and England in the fifteenth where its introduction roughly coincided with the establishment of Caxton's printing press.

The paper on which early maps were printed can be a useful aid in identifying the date of production. Most early maps were printed on paper imported from France or the Netherlands as it was only at the end of the seventeenth century that paper was produced in large quantities in England. During the papermaking process it was customary to impress a pattern into the paper which constituted the manufacturer's *watermark*. A study of watermarks can be helpful in ascertaining the origin and date of printing papers. Since a manufacturer would use a particular watermark over a lengthy period and the paper may have been stored for some time before use the watermark can only provide a rough guide to the date in which a map was printed. It can, however, indicate the earliest possible date of printing.

Although paper has long been the accepted medium for mapmaking cartographers have used a variety of other materials, and still do on occasion; maps have been prepared as tapestries, the county maps of the Sheldon looms in Warwickshire being the most famous example, they have been engraved on gold and silver, produced in marquetry, as embroideries, as frescoes, murals and mosaics. Paper is, however, still the medium which best fulfils the cartographer's requirements of economy, durability and portability.

THE WOODCUT PROCESS

Paper was therefore the ideal medium for the newly-evolved printing processes in fifteenth-century Europe. The earliest method of printing multiple copies was that of taking inked impressions from a woodblock. This is a *relief* method of printing in which the design from which an impression is required is left standing in relief. In the *woodcut* process those areas which were not to be printed were carved away using a flat-bladed knife. Ink was applied to the raised surface, paper brought into contact with it and pressure applied to transfer the inked image to the paper. The timber had to be capable of being easily worked and was sawn from the tree 'on the plank', parallel to the grain. It was then planed to the height of moveable type so that text and illustration could be printed simultaneously. The mapmaker had a

Fig 35 **A manuscript chart of north-western Europe from a portolan atlas (1569) attributed to the Homem family. The artistry achieved by a master of manuscript mapmaking is clearly seen, particularly the beautiful calligraphy and ornamentation. The outline of Britain, however, is extremely inaccurate, even for its period.**
Courtesy: **University of Liverpool   MS.F.4.3**

Fig 36 **Mapmakers have worked in a variety of media. The Sheldon tapestry maps are outstanding works of craftsmanship, particularly in their fine detail of buildings, bridges and other topographic features. This detail is from a tapestry showing parts of Oxford and Berkshire (*c*.1588).**
Courtesy: **Victoria and Albert Museum, London**

choice of three alternative means of transferring his drawn design to the woodblock. He could draw his design directly on to the block; he could draw it on paper, attach the sheet to the block and cut through it; or he could transfer the design with some form of carbon-paper. Little evidence exists to suggest which was most usual. One of the most difficult problems arose from the need to cut the map design in reverse so as to achieve an inked impression which read the right way round. This raised particular problems when much lettering was required, problems which clearly exposed the limitations of the woodcut as a method of map production. The mapmaker was restricted both in the amount and the intricacy of the information he could communicate, and woodcut maps, with notable exceptions such as the world maps of Waldseemüller, generally carried a sparse amount of detail.

One method devised to overcome the difficulties of lettering directly on the block was to set up names in type, cut holes in the block and insert the typeset names. This lent itself to the setting up of titles and legends but the placing and alignment of numerous placenames was a laborious process. *Stereotype* was a more sophisticated technique which involved setting large numbers of names in metal type, making a matrix with some suitable material, then pouring molten tin or lead composition into the mould to form a metal plate. When this had hardened the names were cut individually from the plate and affixed to the woodblock where they were generally placed in the gouged-out areas between the linework of the design. This technique was used in Philip Apian's great map of Bavaria (1568).

The technical difficulties to be overcome in the more sophisticated woodcut maps indicate the use of other tools than the knife. For example it is likely that when executing linework the craftsman would use his knife to outline the design, then remove wood on either side of the lines with a gouge or chisel. The uneven nature of most of the lines produced is evidence enough of the technical problems. Tones were virtually impossible to produce except by very crude shading or stippling and so too were small symbols.

The earliest known woodcut dates from 1418 but the first printed woodcut map was a schematic world map which appeared in a 1472 edition of the *Etymologiae* by the seventh-century Bishop Isidore of Seville. Its impor-

tance in cartographic history derives entirely from its place as the first printed map rather than from the information it conveys. It began a new era in the communication of geographical information which could now be purveyed to a far wider audience than could ever be possible through the medium of manuscript maps.

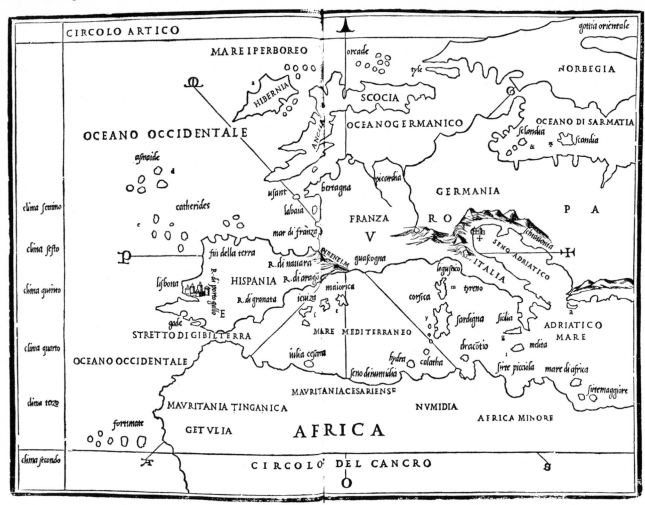

Fig 37 **A particularly crude example of woodcut mapmaking from Benedetto Bordone's *Isolario* (1528). The stylised coastlines are typical of Bordone's work and the map shows little topographic detail except for mountain ranges, a few rivers and a scattering of islands. The only town named is Lisbon though a symbol is located without name on the site of Venice. Climatic zones are named in the left margin.**
Courtesy: **University of Liverpool   Ryl.N.2.30**

The woodcut technique was popular in the fifteenth and sixteenth centuries and woodcut maps can be studied in several Ptolemy editions, particularly the outstanding Ulm edition (1482); in the *Cosmographiae* of Münster and Apian; in the pictorial map of northern Europe by Olaus Magnus (1539); and in numerous lesser works. Two attempts at colour printing have survived as an indication of the desire of early cartographers to enhance the efficiency of the map as a communications medium by developing improved presentation techniques. The first of these was made by Bernardus Sylvanus

in his Venice edition of Ptolemy (1511). Woodward suggests that two printings of the map were taken, one in red, the other in black (*Five centuries of map printing* (Chicago and London, 1975), p. 49). For the first printing, only those names to be printed in red were inked; for the second, the block was either wiped clean or the red names removed, followed by inking of the remainder of the block in black and a second series of impressions taken. The second attempt was more ambitious but less successful. It was an experiment in three-colour printing of the map of Lotharingia in Waldseemüller's Strasburg edition of Ptolemy (1513). Unfortunately the colours lacked consistency and the registration was poor.

A major drawback of the woodcut process was the difficulty of revision, whether it be to correct mistakes or to up-date a block. It was only possible to alter the printing surface by removing sections of the block and inserting fresh pieces of wood which could be cut with the revised detail. The use of inserted metal type or stereotype plates, however, eased the difficulties of revising placenames.

PRINTING FROM ENGRAVED METAL PLATES

The woodcut had limitations both in the size of the map which could be produced and in the quality of reproduction. Linework tended to be rough, lettering was often poorly aligned and tones and symbols were difficult to produce. The process of engraving metal, which had long been practised by Italian gold and silversmiths, offered a superior technique for map reproduction. It was first used for maps in the Bologna Ptolemy (1477) and remained the standard medium for mapmaking for the next three and a half centuries.

Printing from engraved metal plates is an *intaglio* process in which the design elements are incised into the plate by line engraving or etching. In its application to cartography the process, at least up to the 1830s, involved line engraving of copper sheets and it offered several advantages to the mapmaker. First the clarity with which the message was transmitted to the user was improved as a result of the fine detail and precise linework which the craftsman could now produce. Secondly, the difficulties of lettering were minimised and names could be cut in rounded, flowing styles in the plate along with other map detail. Thirdly, the copper plates were not so limited in size as the woodblock had been. Fourthly, alterations were not difficult and revision was a much more practical proposition. As the metal plates were highly durable they could be stored and kept up-to-date over a long period. There were two drawbacks to offset these advantages; first, the printing of text was a relief process while that of maps was intaglio. This meant that text and maps had to be printed separately, using a different press for each; on the other hand, the copper engraving technique was ideal for fairly large single sheet maps. The second drawback was the weight of the metal plates – those used by the Ordnance Survey in making their nineteenth-century six-inch maps, for example, weighing thirty-five pounds.

The engraving process necessitated careful preparation of the plate and its surface had to be highly burnished to a mirror finish. It was then heated and an even coating of wax applied. This was left to harden, before transferring the image from the rough drawing to the white wax coating. The image transfer was achieved either by tracing with a pencil or by placing the drawing face down on the plate, then rubbing the back of it with a burnisher. Occasionally the working drawing was made on transparent paper and transferred by some form of carbon-paper.

The preparation completed, the craftsman then incised his design, usually by line engraving, a skilled technique in which the point of the *burin* removed a small amount of metal to create a fine, shallow channel which would receive the ink before printing. Occasionally an etching technique was used in which the engraver scratched a design through the wax with a needle, then applied acid in order to remove the metal along the lines of the image which was to be reproduced. The etching process was less suitable for mapwork as fewer impressions could be taken from the etched plate. When it was used, the etching process was generally confined to decorative elements of the map such as the title cartouche.

The engraver's basic tool, the burin, was available in two forms; one to cut broad, shallow strokes, the other for the deeper, narrower strokes required to incise intricate detail. Other instruments were at the engraver's disposal as well as the burin. Late-eighteenth-century mapmakers achieved an attractive stipple by rolling a small spiked wheel over the plate. Punches were manufactured for use in producing a series of uniform point symbols such as might be used to locate settlements or industrial sites.

Specialised skills were required for the engraving of different elements of the map design and separate craftsmen were sometimes employed to engrave the outline, lettering, topographical detail and ornamentation.

Corrections to an engraved plate were not difficult. Small blemishes such as scratch marks could be rubbed away with a burnisher and larger areas needing amendment could be smoothed out by hammering the back of the plate. The ease of alteration was advantageous in prolonging the life of a plate and the message it conveyed. This was not necessarily to the advantage of cartography for some publishers used plates so often that detail became so faint as to be barely legible.

The presses used in relief and intaglio printing differed in operation though the principle of transferring an inked image by the application of pressure was the same. In relief printing a flat-bed press was used, the pressure being applied vertically to a sheet of printing paper laid over the type matter or illustration block. The copper plate necessitated the use of a rolling press in which the inked plate and moistened printing paper were forced smoothly between two rollers which progressively forced the printing ink from the incised lines on the plate to transfer the image to the paper.

Many cartographers were equally skilled as engravers but others laid out the geographical data in map form then prepared a fair drawing for the skilled engraver who would incise the plate ready for printing.

Copper plates were capable of a very long life. Skelton suggests that 2000 to 3000 impressions could be taken from the same plate without serious deterioration but with the exercise of care and judicious retouching its life could be further extended (*Decorative printed maps of the fifteenth to eighteenth centuries* (London, 1965), p. 3). Plates of Speed's county maps, for example, were used many times between the first publication of his *Theatre* in 1611 and their final reprinting by C. Dicey & Co. in 1770. Clearly, however, the use of plates over such a long period was undesirable as the message became progressively more out-dated.

Copper engraving remained in favour for map reproduction up to the middle of the nineteenth century when the advantages of lithography, later combined with photographic processes, became obvious. Steel engraving had some devotees, the harder metal making very long printing runs possible without undue wear on the plate and without the introduction of new machinery. This very hardness, however, made the process of engraving the plate more difficult and maps engraved on steel were the exception rather than the rule.

Fig 38 **Tools used by engravers in copper; a rolling press for printing copper plates such as maps; a map of Savoy by J. Jansson which illustrates the elegant, delicate work which could be produced by master craftsmen.**

HAND COLOURING

Prior to the introduction of colour printing in the nineteenth century, maps could only be coloured by hand and were sold with or without colour according to a customer's wishes. The extravagantly-ornamented maps of the sixteenth and seventeenth century benefited considerably when hand-colouring was carefully applied by a skilled illuminator but unfortunately many maps of the period were coloured so crudely that detail was obscured and the quality of the engraving impaired. The skill with which colouring took place could be one of the determining factors behind successful communication. By picking out and differentiating features of the map, colour facilitated map reading. How much easier it is, for example, to follow the course of boundaries when outlined in colour or to examine the distribution of settlement or woodland. The profession of colourist was highly respected. Ortelius, for instance, started his distinguished career as an illuminator of maps and the Frenchman, Nicholas Berey, enjoyed the title of *enlumineur de la reine*.

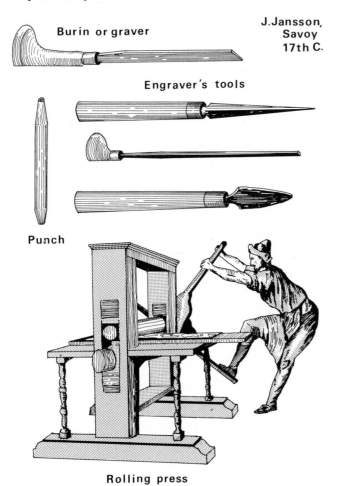

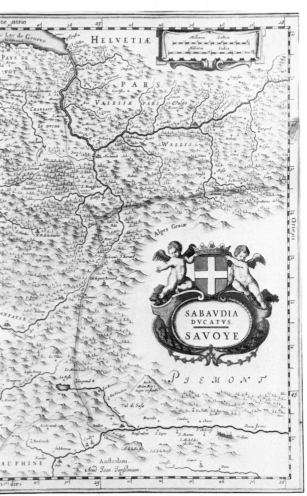

LITHOGRAPHY AND MAP PRODUCTION

Lithography (literally 'drawing on stone') is a *surface* or *planographic* process in which the areas which print and those which do not are all on the same level. The process depends on the antipathy between grease and water and on the affinity of one greasy substance for another. It was discovered in 1796 by a Bavarian, Alois Senefelder, who experimented with 'do-it-yourself' printing in an attempt to find an economical means of reproducing his literary works. After experiments with copper plates and stereotyping he tried a smooth, porous limestone which was quarried at Kellheim in Bavaria. He manufactured a greasy ink composed of oil, wax and lamp-black and quite by chance wrote on the stone with the intention of subsequently copying his writing on to paper. He then pondered on the feasibility of etching the stone with acid so that the writing would be left in relief ready for printing. From this he discovered that the stone retained any writing made on it with his greasy ink and that if he wetted the stone, then inked it, the ink would adhere only where his original greasy ink writing had been made. If he then pressed paper on to the stone it picked up the inked marks while the stone retained the image for an indefinite number of printings. Senefelder went on to develop a lithographic printing press with which he claimed to run off some 1200 prints a day. In the early nineteenth century he became involved with map printing and in 1809 was appointed Superintendent of Lithographic Printing to the Bavarian Cadastral Survey.

Interest in lithography spread outside Germany. In Paris, Edme-François Jomard, curator of maps in the Bibliothèque Nationale, was especially quick to see lithography as a means of making facsimiles of maps in his care readily available to students everywhere. Lithographic reproductions were therefore made from hand copies of selected maps and published in atlas form. By 1820 lithographic presses were in use in Austria, Belgium, Italy, the Netherlands, Spain and Russia and the new process was put to military needs, reproducing official documents, in France in 1818. In Britain a press was set up at the Quartermaster General's Office in Whitehall in 1807 and one of its first cartographic products was a plan of Bantry Bay dated 1808. By 1819 this press had produced numerous maps and plans related to the Peninsula War in addition to maps for the control of home-based forces in Britain. The operation of the Quartermaster General's press deserves a place in cartographic history as one of the earliest uses of lithography for purposes of military mapping.

In 1849 Manuel Francisco de Barros, Viscomte de Santarem, a Portuguese historian, published a facsimile atlas *Atlas Composé de Mappemondes, de Portulans et de Cartes Hydrographiques et Historiques* in which the maps, like those of Jomard, were lithographed from hand-drawn copies. Santarem chose lithography rather than copper engraving partly because of the excessive cost of the latter, partly because of the ease with which the artist's drawing could be reasonably faithfully transferred to the stone. Although the facsimiles of Jomard and Santarem were copies and subject to human errors, their work made a notable contribution to the systematic study of cartographic history by making the more important early maps available to researchers throughout the world.

In 1818 and 1819 French, German and English editions appeared of Senefelder's *Complete Course of Lithography*, a work which did much to further the spread of the technique. Between 1820 and 1850 lithography was used in Germany for varied cartographic purposes – topographic series, urban plans, wall maps and atlases. Two-colour maps were successfully printed at Freiburg in the early 1830s. In France, too, lithography developed rapidly but in Britain the Ordnance Survey and the Admiralty continued to rely on copper engraving throughout the first half of the nineteenth century.

Lithography was enthusiastically welcomed in the USA and a map of the Catskill Mountains illustrating geological strata was issued in 1822. The early period of American lithography was also that of the canal and railroad, and the evolving transport systems increased the demand for maps, many of which were printed by lithography. Official mapping agencies, though still in the early stages of development, made some use of the new process but continued with engraving for much of their output. Lithography was also tried for military mapmaking but the large lithographic stones were awkward for field use.

THE INTRODUCTION OF PHOTOGRAPHY TO MAP REPRODUCTION

The application of photography to the map reproduction process caused a revolution in the ability of maps to communicate information rapidly to a wider public. Experiments were made in 1855 at the British Ordnance Survey under the direction of Colonel Sir Henry James and at roughly the same time photography was applied to map production in Australia, the United States, and Belgium. With the aid of the camera it was now possible to photograph a fair drawing made in dense black ink and the laborious processes of engraving lines on metal plates or lithographic drawing on stone became

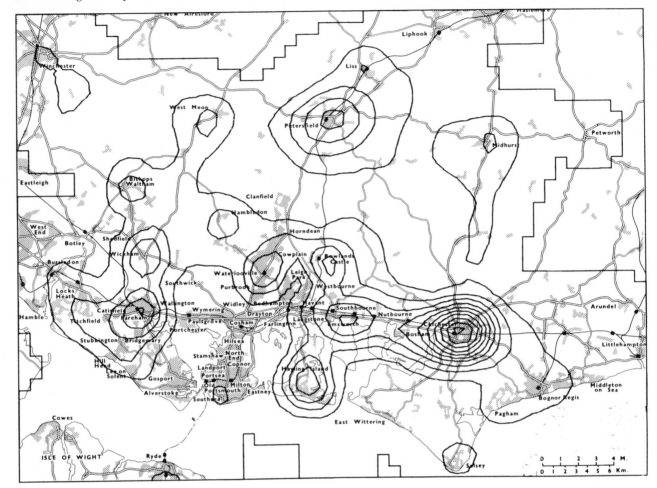

Fig 39 **Isolines prepared by computer and graph plotter on pre-printed base map to show areas in which a sample of Portsmouth dwellers would prefer to live. The closer the concentration of isolines the greater the preference for a particular area; Chichester, therefore is clearly the most desirable locality.**
Courtesy: **Alan Duffett**

unnecessary. Maps could be reproduced in faithful facsimile, one of the resulting publications being the invaluable *Facsimile-Atlas* (1889) of A. E. Nordenskiöld which reproduced important fifteenth- and sixteenth-century maps and introduced a systematic approach to the study of early cartography.

One problem which remained to be solved was that of finding a lightweight printing surface to replace the cumbersome lithographic stone. *Zincography* was one solution. In this process printing took place from a zinc plate on which a map had been drawn with lithographic ink or transferred from an original drawing. *Photozincography*, in which the image was transferred to the plate by photography, was used by the Ordnance Survey *c*.1860.

WAX ENGRAVING

Cerography or wax engraving was invented by Sydney Edwards Morse who published a *Cerographic Atlas* in 1841. The process, which became a popular method of making maps in the USA, involved spreading a thin layer of wax over a polished plate, usually copper, then transferred an image to the wax by draughting or photography. This design was then engraved on the wax and lettering applied by pressing printer's type into it. After engraving was complete the plate had to be built up by applying more wax between the engraved lines so as to ensure that the plate would be deep enough for printing. The plate was then put into an electrotyping bath and a shell of copper deposited on the wax. The copper shell was removed and a layer of molten type metal poured over the back of it. The resulting plate was planed and mounted on wood to type height. American maps printed by cerography had a mechanical, over-lettered appearance.

THE OFFSET PRINTING PRESS

In 1904 an American, Ira W. Rudel, invented the offset printing press to bring about another basic change in map reproduction. The principle of the offset press was that printing surface and printing paper were not in direct contact, the inked image being first offset on to a resilient rubber blanket on another cylinder. The offset press resulted in greatly increased production speed. Early presses were able to run off some fifteen hundred prints per hour while those in use today can achieve five times that speed.

RECENT DEVELOPMENTS

Since the end of World War II cartographic techniques have changed with startling rapidity. The use of plastic materials as a drawing medium, the introduction of lettering by photo-composition and the dry-transfer system, are among the many innovations which combine to produce clearer, more legible maps. The draughtsman himself has become a new kind of engraver, producing images by *scribing* i.e. cutting lines and symbols in plastic sheets of great dimensional stability. Scribing produces cleaner, more consistent linework than ink drawing and the use of dimensionally stable plastics ensures accurate registration, or fit, of colours.

The most significant development, however, has been the application of digital computers to form a graphic image in map form. At this point a distinction must be drawn between 'computer aided cartography' and 'automated cartography'. The former normally indicates a process of producing thematic maps while the latter denotes the use of a computer in the production of conventional topographic maps. In thematic mapmaking the computer is advantageous for the rapid processing of data and its speedy transformation into map form. Three facilities of visual presentation are available; first, by means of a *line printer*, a process in which the maps produced are composed of characters similar to those of a typewriter, symbols being produced from single characters or by overprinting two or more; secondly, by using a *graph plotter* or a *drum plotter*, devices in which a pen is activated to produce a series of straight lines on a roll of paper, the pen moving on a carriage along a horizontal axis while the paper unrolls along a vertical axis. While the resulting map is made up of jerky lines it bears more resemblance to the conventional map than the products of a line printer; thirdly, by presenting a 'temporary' map by cathode ray tube, the image appearing on a screen similar to that of a television set.

The basis of computer mapping is a system of horizontal and vertical co-ordinates from which the computer can construct an outline base map. Automatic devices known as *digitizing tables* allow the co-ordinates to be quickly determined and punched on to magnetic tape. The co-ordinates of all data points are determined and punched in the same manner. Their values are punched separately and all the information is fed into the computer which transforms it into the requisite form. The future progress of computer mapping will depend to some extent on the speed with which *data banks*, in which information is stored on tape, can be made available. Data extracted from such banks can then be used in conjunction with a suitable computer mapping program to produce thematic maps using one of the standard statistical presentation techniques. Several packages are available; the SYMAP program produces isoline maps (those based on lines along which values are constant, e.g. contour lines) and choropleth maps (those showing density per unit of area, e.g. number of persons per square kilometre); LINMAP, developed by the Ministry of Housing and Local Government in Britain, was intended for use by planners in mapping census data; COLMAP permits the use of colour in thematic mapmaking; MAPIT is designed to produce flow maps, dot maps and graduated symbols. While a great asset of the computer is its speed of production, it offers another facility which is of great value to the map compiler. This is the possibility of speedily presenting several different versions of the same map, perhaps at varying scales, on transformed projections, or using different statistical presentation techniques. From the resulting maps the compiler can select those which most closely fulfil his requirements.

The application of the computer to the production of topographic maps has been less spectacular and the automation of the mapmaking processes is far from complete. It should be stressed also that although the use of computers in any form of mapmaking introduces a completely revolutionary concept, the basic cartographic processes remain unchanged. The new technology has simply been applied to use these processes more quickly and with greater flexibility. Nevertheless, it has been suggested that computer mapmaking will have replaced manually produced maps for over ninety per cent of all maps produced by AD 2000. Many of these maps, however, may be of the 'temporary' variety viewed on a display screen. As far as map content and format is concerned there may be restrictions on the ranges of symbols, on the styles of lettering and on the size of maps. To offset these limitations there will be greater flexibility in the ease of projection construction, the ease of presentation at different scales and the rapidity of data processing.

Certainly it is essential that cartographers accept the rapid developments which are taking place and come to terms with them. It is an intriguing thought that fifty years hence the sheet map as we know it may be just as much a part of cartographic history as the clay tablet of the Babylonians.

5 THE EVOLVING WORLD MAP

Scientists when attempting to trace the origins of a particular human activity often lack written evidence or tangible relics and in such circumstances it is customary to examine the corresponding activities of pre-literate peoples in recent times and to draw conclusions by analogy. Such a procedure has been applied to cartography.

It is well known that human beings have an innate ability to prepare rudimentary maps – how much easier it is, for example, to explain a recommended route to a stranger by means of a simple sketch map drawn on the back of an envelope than to do so verbally. In the same vein, explorers have commented that native peoples in various parts of the world will show the way to a certain place by drawing a sketch in the soil or sand with a stick. The remarkable mapmaking activities of Eskimo peoples using wood as their main medium have been well documented. So too have been the sea charts made by the Marshall Islanders who used shells to represent the location of islands and the ribs of palm leaves to indicate the direction of wave fronts. If such peoples are able to make recognisable maps it can be assumed that the peoples of antiquity who enjoyed a high degree of civilisation and culture also practised mapmaking. Although hard evidence is fragmentary it is believed that the idea of recording information about the environment in map form was conceived in the Middle East by the peoples of the fertile river valleys of the Nile, Tigris and Euphrates. The early Babylonians made maps at varying scales from estate plans for cadastral purposes to small subjective illustrations of their concept of the universe. Some of their work has survived, for the medium in which they chose to draw their maps was the baked clay tablet, the design being first incised into the clay while wet, then baked to give durability. Such a medium lacked most of the properties desirable for mapmaking, especially for the map to be an effective means of communication. The tablets were difficult to work on, too small to allow much freedom of expression and yet lacked portability. Throughout history man has constantly sought to provide a graphic display of his speculations about the

nature of the earth and its place within the universe, and a tiny Babylonian clay tablet, dating from *c.* 500 BC, is regarded as the earliest known example of such a display. It illustrates, a round, flat earth encircled by ocean, beyond which are located the seven islands thought to form a bridge to the outer circle or Heavenly Ocean. It is centred on Babylon with the Euphrates seen flowing southwards from the Armenian Mountains to the Persian Gulf. The originator of the map supplemented the

Fig 40 **Babylonian world map (*c.* 500 BC), a schematic representation of the universe, centred on Babylon, and incised into a clay tablet.**
Courtesy: **British Library Board, Department of Western Asiatic Antiquities 022791**

graphic portrayal with a verbal explanation in cuneiform characters of the map's astrological and religious significance.

Cartographic evidence from the ancient world is sparse, particularly from Egypt. Despite the advanced level of Egyptian civilisation and the data available to Egyptian scholars from military campaigns, trading missions and general exploratory journeys, nothing remains to supply cartographic proof of Egyptian speculation about the earth as a whole and about the universe.

THE GREEK CONTRIBUTION; THE BIRTH OF SCIENTIFIC CARTOGRAPHY

Sufficient evidence exists to indicate that the foundations of scientific mapmaking were laid in Ancient Greece. Indeed it may be said that the Greek contribution was more advanced than most mapmaking prior to the fifteenth century.

Greek mapmaking of a routine practical nature was undertaken by land measurers or *geometers* who were engaged in surveying small areas only, but it was a theorising about the nature of the earth and its representation in map form which dominated early Greek geography. Greek scholars pondered about the physical shape of the earth and at various times saw it as a flat disc, a cylinder or a globe. Anaximander is thought to have constructed a map of his known world *c.* 550 BC and fifty years later improvements were made to it by Hecataeus. The habitable world, or *oekumene*, as the Greeks knew it extended from the Atlantic Ocean to the Indus, was roughly rectangular in shape and twice as long from east to west as from north to south. By the fourth century BC Greek philosophers were investigating the revolutionary concept of a spherical earth, basing this on a mixture of scientific observation and the idea that the earth, as the master creation of the Gods, must have the perfect shape, a sphere. Once the sphere became accepted scholars took a natural interest in its dimensions. Eratosthenes (276–196 BC), head of the Library at Alexandria, achieved a remarkably accurate determination of its circumference, his figure of *c.* 28,000 miles being correct to within fourteen per cent. With the establishment by this time of chorography, geography and cosmography as the studies respectively of regions of the earth, of the earth as a whole and of the universe, scholars saw the necessity of portraying the results of their studies graphically in map form. Their belief in a spherical earth led them to see that geographical locations could be displayed within a geometrical framework. In this context Eratosthenes recognised the need to formulate some form of reference system by which places could be located on the globe and divided

the *oekumene* into northern and southern parts by drawing an east-west line through the Pillars of Hercules and the island of Rhodes, then a north-south line passing through Alexandria, Rhodes and the Dardanelles. These two lines, designated the Parallel of Rhodes and the Meridian of Alexandria, formed the basis of a system of irregularly-spaced parallels and meridians. Eratosthenes thus pioneered the grid reference system which remains a basic means of defining position and he paved the way for future attempts at mapping the earth's curved surface on a plane sheet of paper. All was not plain sailing, however, for the problem at once arose of accommodating the small area of the known world into the vast expanse of the sphere. The philosopher Crates attempted to overcome this difficulty in 150 BC by constructing a globe on which he drew three additional continents to balance the *oekumene* and in so doing it could be said that he anticipated the discovery of the Americas and Australasia.

Scholars were now aware of a physical environment which lay outside their own personal experience and were attempting to express their new view of the world in carefully constructed maps based on a network of parallels and meridians with the interior detail completed from information supplied by the itineraries of contemporary travellers. At this point a colossus of the geographical world emerged in the person of Claudius Ptolemy (AD 90–168), Librarian of Alexandria. Ptolemy was pre-eminent as a geographer and astronomer, expounding his astronomical theories in *Almagest*, a work which was unchallenged in its field for centuries, and his geographical knowledge and beliefs in *Geographia*, a treatise which is one of the great landmarks in cartographical development and which had resounding effects on Renaissance mapmaking some fourteen hundred years after it was written. In the Alexandria Library the earlier studies of Eratosthenes, Marinus of Tyre and other Greek philosophers were available to Ptolemy, and by adding information gleaned from travellers, navigators and merchants visiting Alexandria he was able to assemble the accumulated geographical knowledge of his day in one impressive treatise. *Geographia* contained all the ingredients from which maps could be constructed – the theoretical principles of cartography; a discourse on mathematical geography; instructions for constructing map projections (in which he showed his awareness of the inevitability of some deformation whichever type of projection was used); a list of some 8000 placenames with their latitudes and longitudes. Ptolemy therefore made major contributions to the developing cartographic language and using his data and his detailed instructions it was now possible to depict the world in a scientific framework, locating

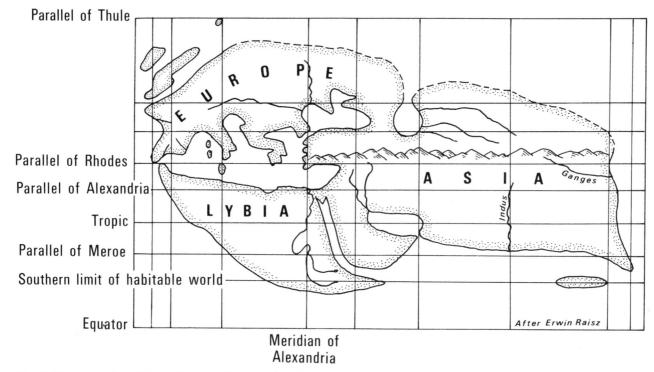

Parallel of Thule

EUROPE

Parallel of Rhodes

Parallel of Alexandria

Tropic

Parallel of Meroe

Southern limit of habitable world

Equator

LYBIA

ASIA

Ganges

Indus

After Erwin Raisz

Meridian of
Alexandria

Fig 41 Reconstruction of Eratosthenes' reference grid with his system of irregularly-spaced parallels and meridians based on the Parallel of Rhodes and the Meridian of Alexandria.

places by means of their geographical co-ordinates. Ptolemy also made some fundamental errors which had repercussions in European mapmaking until 1700, the most significant being his underestimation of the earth's circumference and his estimate of the Mediterranean's length to be 62° rather than the true 42°.

THE CARTOGRAPHY OF IMPERIAL ROME

Visible proof of the practical nature of Roman mapmaking is seen in route maps such as the *Peutinger Table* and some idea of their wider thinking is provided in literary sources and in the schematic map/diagrams which illustrated medieval editions of classical texts by scholars such as Sallust and Macrobius. There are also references to a world map by Marcus Vipsanius Agrippa (63–12 BC) which is believed to have been based on military road surveys within the Roman Empire. The map was completed in 20 BC and after Agrippa's death it was publicly displayed in the Campus Martius in Rome. Although copies were distributed to principal stations of the Empire none has survived, though there have been suggestions that the *Peutinger Table* could be a modified version of Agrippa's work. The Romans generally lacked the intellectual concern of the Greeks for geography and were more interested in the practical aspect of

determining the distances between places than in establishing their geographical co-ordinates. Nor were they particularly concerned with speculation about the physical nature of the earth and merely followed the Babylonian theory of a flat disc of land encircled by sea.

ISLAMIC CARTOGRAPHY

Following the dissolution of the Roman Empire, cartography in Western Europe went into a period of decline with scant regard for early scholarship. In the Islamic world, however, Greek traditions of learning and scholarship were maintained. Arab geographers had access to manuscripts of Ptolemy's *Geographia* which had been lost to the west and in the eighth century AD the work was translated into Arabic. To add to their geographical accomplishments, Arab scholars were skilled in astronomy, mathematics and geometry, and it is not surprising that Arabic cartography followed the Greek pattern, though Arab geographers made distinctive contributions of their own. Ptolemy's work was respected but not slavishly followed – one of the important features of Islamic cartography was the determination of latitude and longitude of places by astronomical observation, and Ptolemy's erroneous length of the Mediterranean was reduced in the ninth century by al-Khwarizimi to 52° and as a result of observations by al-Zargali in the twelfth century to the correct 42°.

From the eighth to the eleventh centuries there was a strong tradition of Arabic religious cosmography and

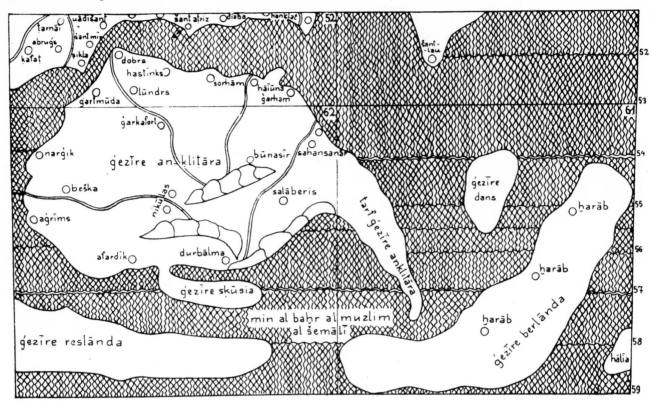

Fig 42 **Detail from a re-construction of the Al-Idrisi world map (12th century) showing the British Isles.**
Courtesy: **British Library Board   MAPS REF 04b P145 PL30**

numerous treatises were illustrated by small circular wheel maps or T-O maps, similar in design to those being prepared in Western Europe – a circular, flat earth encircled by sea (representing the 'O', the Mediterranean representing the vertical stroke of the 'T' and a line from the Don to the Nile its horizontal bar). Unlike the religious maps of the west, however, which were centred on Jerusalem, those of the Islamic world were not unnaturally centred on Mecca. The world map by al-Istakhri (934) is representative of the *genre*. It is a very stylised portrayal with a highly distorted view of the continents of Africa (its southern regions pointing to the east), Europe (seen as an island) and Asia. Other Arabic wheel maps were divided into latitudinal climatic zones with no geographical features indicated.

The most significant Arabic map was the world map of al-Idrisi, an Arab scholar born at Ceuta in 1109 who became geographer to the Norman King Roger of Sicily at Palermo. Al-Idrisi compiled a book which assembled all currently available data on the latitudes and longitudes of places and the distances separating them. His map also showed places within their appropriate climatic zone, for al-Idrisi used nine parallels of

latitude based on climate and eleven meridians of longitude. He made no allowance for the earth's curvature. A variety of sources were used in compiling the map which combined Ptolemaic theories and up-to-date information gleaned from al-Idrisi's own travels and other Islamic itineraries.

THE MEDIEVAL PERIOD

In the western world cartography was virtually extinguished as a scientific study during the Dark Ages and world maps produced at this time were merely vehicles for subjective speculation and the graphic display of theological teachings rather than the outcome of systematic reasoning. The medieval world view was expressed in the writings of Isidore, a seventh-century Bishop of Seville who declaimed, 'the earth (Orbis) was named from its roundness . . . Europe and Africa were made in two parts because the great sea (called the Mediterranean) enters from the ocean between them and cuts them in half'. Such beliefs were cleary demonstrated in the typical maps of the time, the T-O or T in O maps, a designation which reflects their layout, a similar one to that of the Arabic wheel maps already mentioned except for Jerusalem, rather than Mecca, being placed at the centre. Many manuscripts dating from the eighth to the fifteenth centuries contained this type of map, the authors' sole concern being to provide a graphic por-

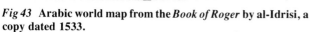

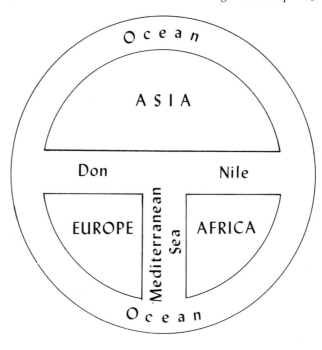

Fig 43 Arabic world map from the *Book of Roger* by al-Idrisi, a copy dated 1533.
Courtesy: Curators of the Bodleian Library   MS. Pococke 375 fols. 3v–4r

Fig 44 Diagram showing the principle of the T-O or T in O map with the tri-partite division of the world between the sons of Noah.

trayal of their religious ideas which divided the world between Ham, Shem and Japheth, the sons of Noah.

An alternative form of world map in vogue during the medieval period was the hemisphere or climatic zonal map. Macrobius, a Greek scholar whose work enjoyed great popularity, used this type of map extensively to illustrate his texts, dividing the world into seven climatic zones, a theory based on early Greek science, developed as already mentioned in the Islamic world, and which caused no offence to the powerful European theologians.

From the eighth century onwards maps were produced in great numbers, mostly theological in concept. Among the outstanding cartography of the period was a group of ten world maps based on an original design by a Spanish priest, Beatus, in the eighth century, whose text, *Commentary on the Apocalypse* (*c.* AD 776) included a world map, now lost, which appears to have been compiled to demonstrate the teachings of the Old and New Testaments and to convey graphically the spread of the Christian faith through the habitable world. This was achieved by placing a representation of each of the Twelve Apostles in a location traditionally associated with his preachings. Ten maps derived from the original Beatus map survive, varying in date from the tenth to the

thirteenth centuries. All occur in manuscripts of the *Apocalypse Commentary*. The finest was made in the Aquitanian convent of St Sever (*c.* 1030) and is thought to be nearest to the original in design and content. It is an oval map encompassing two pages, the continents roughly disposed in the T-O pattern with the encircling ocean filled with pictorial illustrations of fish and boats. The artist managed to convey a good deal of topographical information, showing mountain ranges and rivers prominently and symbolising major towns by a building seen in elevation. He gave particular prominence to Rome and to St Sever, clearly wishing to indicate the latter's importance as the place where the map was made.

With their comparative wealth of topographical detail and the absence of features based entirely on fantasy, the Beatus-based maps provide a useful link between the ancient and medieval worlds.

The Cottonian or Anglo-Saxon Map *c.* 1000

This world map found in a Latin manuscript of Priscian's *Periegesis* is one of the more interesting world portrayals for it shows with a rare degree of accuracy places and features not illustrated elsewhere and in its coastal

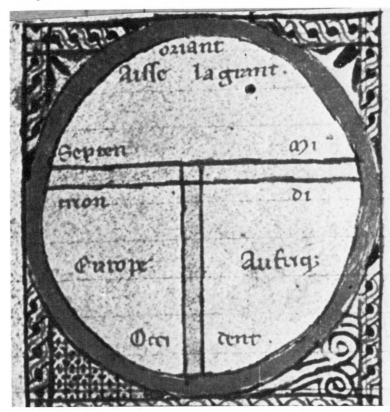

Fig 45 Early 14th-century T-O map by Brunetto Latini.
Courtesy: **Curators of the Bodleian Library   MS Douce 319, f.iii of contents table (P & A ii, no.154)**

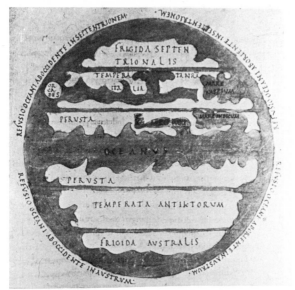

Fig 46 An alternative concept to the T-O map was the zonal map in which the world was displayed as a single hemisphere with encircling ocean and divided into climatic zones. In this example by Macrobius, made in the 11th century, a further ocean divides the hemisphere into two parts.
Courtesy: **Curators of the Bodleian Library   MS.D'Orville 77, folio 100**

delineation surpasses any map prepared before the fourteenth century. Drawn within a rectangular frame rather than the more usual circle, with east at the top, the map owes no particular allegiance to any of the standard types of medieval map. Beazley suggested that the map owes part of its excellence to being the work of an Irish monastic scribe trained in the great Irish schools of learning, art and science(*The dawn of modern geography*, vol. 2 (New York, 1949), pp. 561–2). Certainly the map's delineation of Ireland is superior to that of other medieval world maps and it was the first map to mark an Irish city – Armagh. Beazley believes that it was made specifically to illustrate the settlement of the Twelve Tribes of Israel but adds that a Roman provincial map may have been the source of data about Asia Minor, central and south-eastern Europe and North Africa. In modern terminology it could be described as a thematic map if Beazley's theory is correct.

EARLY CHINESE CARTOGRAPHY

While scientific mapmaking in Europe, after its development by the Ancient Greeks, was interrupted for a millennium by the dominance of cartography based on theological beliefs, that of China showed no such interruption and during the period when the T-O map

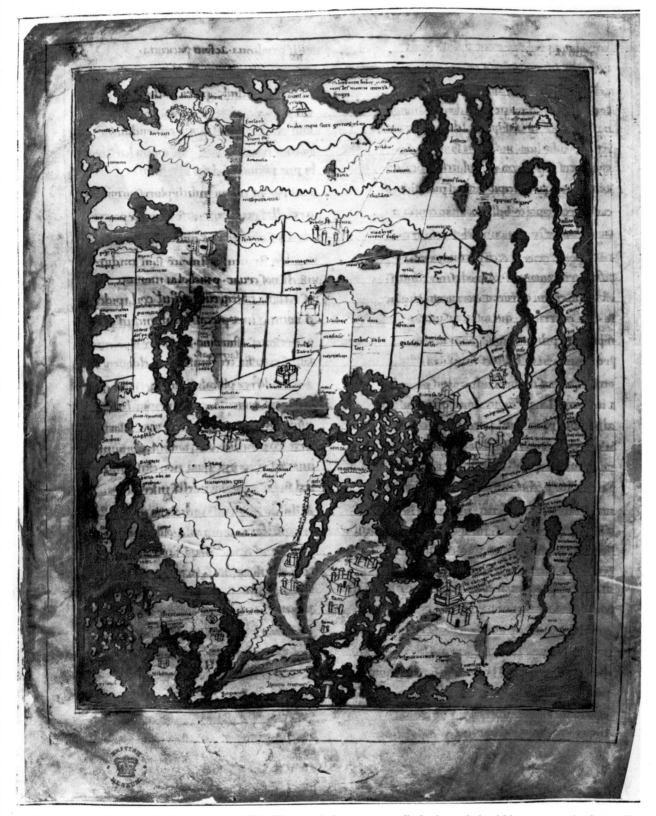

Fig 47 The Cottonian or Anglo-Saxon map (*c.*1000). The map is drawn, unusually for its period, within a rectangular frame. The depiction of the British Isles is remarkably good for its date and it has been suggested that the map may have been the work of an Irish monk.

Courtesy: British Library Board   Cotton MS.Tib.B.V.fol.56V

was the typical form of European cartography, Chinese mapmakers were developing a scientific tradition of their own. Two figures emerge as being the equal in importance in the east of Ptolemy and Eratosthenes in the west. The astronomer Chang Heng is believed to have originated the rectangular grid system which is such a characteristic feature of early Chinese cartography. In AD 267 Phei Hsiu was appointed Minister of Works. In the course of his work he studied the Yü Kung (the oldest surviving Chinese geographical document) and found that many changes had taken place in the names of places, rivers and mountains given in the Yü Kung. As a result he classified the ancient names as far as possible and made a geographical map in eighteen sheets from the information in the Yü Kung. In the preface to this map Phei Hsiu wrote that

> In making a map there are six principles observable; (1) The graduated divisions, which are the means of determining the scale to which the map is to be drawn. (2) The rectangular grid . . . which is the way of depicting the correct relations between the various parts of the map. (3) Pacing out the sides of right-angled triangles, which is the way of fixing the lengths of derived distances (i.e. the third side of the triangle which cannot be walked over). (4) Measuring the high and the low. (5) Measuring right angles and acute angles. (6) Measuring curves and straight lines. These three principles are used according to the nature of the terrain and are the means by which one reduces what are really plains and hills to distances on a plane surface.

Phei Hsiu went on to describe the use of the graduated divisions of the rectangular grid in achieving uniform scale throughout the map. Like Ptolemy therefore, Phei Hsiu provided clear instructions for mapmaking and though none of his maps survive, the principles he outlined were followed by later Chinese mapmakers.

THE GREAT EUROPEAN 'MAPPAE MUNDI'

Although the majority of the theologically-orientated European maps of the medieval period were conceived as tiny illustrations to monastic texts, a few were extremely large and detailed and were independent works in their own right. These large maps were essential adjuncts to any monastic library from the ninth century onwards and examples are known to have been housed at St Gall, Reichenau, Tegernsee and Bamberg. The only surviving example is the remarkable map made c.1300 and housed in Hereford Cathedral. An equally fine and even larger map (thirteen feet in diameter as against the five of Hereford) was housed in the German monastery of Ebstorf but was a casualty of World War II. The two maps were similar in principle, each being a modified version of the T-O design, conceived in the context of Christian theology and summarising the geographical thinking, both secular and religious, of the Middle Ages. The Christian symbolism is immediately obvious; on the Hereford Map the figure of Christ presides as at the Day of Judgement at the top of the map (very much as he is portrayed in the *tympani* of many medieval churches); the Garden of Eden is placed in the east and biblical references abound – Jerusalem and Calvary at the centre, the Tower of Babel, Noah's Ark and many more. The Hereford Map was prepared on a calf-skin measuring sixty-five inches by fifty-three and drawn with a quill pen. The map bears no signature but an inscription identifies the author as Richard of Haldingham. There are alternative theories as to the map's *raison d'être*; one school of thought suggests that it was intended to serve as a *reredos* or background for an altar; the other sees it as designed purely for educational purposes to communicate the teachings of the Christian faith by pictorial means. The people of the Middle Ages, untutored and illiterate, viewed the world with a mixed fear, faith and suspicion. They were familiar, nevertheless, with medieval pictorial stained glass, wall paintings and sculpture, and could comprehend the graphic display of the world in map form. From the map, itself virtually a pictorial encyclopedia, they assimilated smatterings of natural history, geography and classical mythology in addition to the teachings of the Christian faith. The Bible was the primary source of data for the map, other detail being derived from classical texts, earlier maps, Roman itineraries, pilgrim and trade routes, medieval histories and monastic bestiaries. Its contemporary value as a means of communicating information to a wide audience can be assessed from the knowledge that, when it was made, thousands of pilgrims were visiting Hereford Cathedral daily to see the shrine of Bishop Cantelupe. These visitors must have marvelled in the picture of the world provided by the map and gone away to spread far and wide the story of its lands, rivers, biblical lore, fantastic creatures and mythical beings like the man shading himself from the sun under one large foot.

The great *mappae mundi* therefore were an effective means of disseminating theological teachings. As cartographic works they were decorative and full of fascination but unreliable in their content, based as it was on a combination of fact and pure fantasy.

The world map of Fra Mauro (1457–9)

The zenith of monastic mapmaking was reached with a large circular world map made to a commission of King Afonso of Portugal by Fra Mauro, a monk cum professional cartographer from Murano, an island in the Venetian lagoon. Over six feet in diameter, the map surpassed

Fig 48 The Hereford Map, made c.1300 by Richard of Haldingham, is the most remarkable surviving large medieval world map. The British Isles are grossly distorted in the bottom left to fit into the circular frame and their delineation does not bear comparison with that of the earlier Cottonian map.
Courtesy: **The Dean and Chapter of Hereford Cathedral**

in accuracy and detail anything so far produced. The amount of detail is considerable – indeed, Bagrow regarded it as an over-abundance in which the important accurately-delineated geographical features are intermingled with superficial data based only on hearsay (*History of cartography* (London, 1964), p. 73). In the execution of his map, Fra Mauro worked along with a chartmaker, Andrea Bianco, and a team of assistant draughtsmen and illuminators. The Murano monastery's accounts contain records of payments made to these men and also for materials used in making the map. The content of the map owes something to contemporary marine charts, something to Ptolemy (some of whose misconceptions – that of a landlocked Indian Ocean, for example – were rectified by Fra Mauro) and much to information supplied by travellers' itineraries, including the writings of Marco Polo. Fra Mauro made a clear break with ecclesiastical tradition by neither centring his map on Jerusalem nor placing east at the top. At first glance the unusual orientation with south to the top of the map makes its message difficult to assimilate, particularly as the most accurate and immediately recognisable part of the map is the European continent. Compression into a circular format has made some distortion inevitable and the configuration of lands in the Far East and of Africa south of the Equator is decidedly vague.

This great map marked the end of cartography dominated by theologians and a new era of scientific mapmaking was about to begin.

THE CARTOGRAPHY OF RENAISSANCE EUROPE

At no other period in history did man's conception of the world change more quickly than it did around the year 1500. Cartography, like other arts and sciences, underwent a major rebirth and scholars turned away from theologically-orientated maps back to the scientific legacy of the classical world.

Three major factors contributed to this surge of scientific cartography: first, manuscripts of Ptolemy's *Geographia* were reintroduced into Western Europe and the repossession of these manuscripts, followed by their speedy translation into Latin, provided a stimulus to contemporary cartographers. During the fifteenth century, *Geographia* was superbly transcribed in manuscript before the introduction of printing brought about a spate of printed editions, the production of which caused an odd situation in which the maps which showed the world and its regions as seen by Ptolemy many centuries earlier were being reproduced by the latest technological processes.

Particularly representative of Ptolemaic geography is the fine world map on a modified spherical projection which appears in the Ulm edition of *Geographia* (1482) edited by Nicolaus Germanus, a Benedictine monk of Saxony. This is a handsome woodcut map which demonstrates the fallacies of Ptolemaic theory as well as its good points: the Indian Ocean appears to be completely landlocked; the Indian peninsula is so truncated as to be virtually non-existent; a large island (Taprobana) is placed in the location of Ceylon; the east-west extent of the Mediterranean Sea is exaggerated; Scotland appears at 90° to the rest of the mainland. These are some of the 'trademarks' by which Ptolemy-derived maps are instantly recognisable. The Ulm world map is surrounded by twelve wind-blowing heads which are an elaboration of the ancient classical concept of four wind directions and are used by Ptolemy as an indicator of direction. Considerable topographical detail appears on the map with rivers and mountain ranges extremely prominent. The map is covered by a network of meridians and parallels and the left-hand margin is subdivided to show the apportioning of the world into climatic zones.

The second factor which influenced the cartographical revival was the application of the newly-developed craft of printing from engraved surfaces to mapmaking (newly-developed, that is, as far as Europe was concerned, for printing had been applied to Chinese mapmaking in the twelfth century, the first map to be printed in China dating from *c.*1155). Multiple impressions of maps were now possible, each faithfully reproducing the original engraving, so that the human errors of the copyist were obviated. This new development, one of the most momentous in cartographic history, revolutionised the map as a communications medium, for maps could now be made available to a wider (though still not very extensive) audience rather than a minority group of scholars, statesmen and rich merchants. It made possible the publication of many printed editions of Ptolemy, some with woodcut maps, others using the more refined process of printing maps from engraved copper sheets.

If the introduction of printing meant much in the field of cartographic production, the content of maps was equally affected by the third factor contributing to the new concept of mapmaking, i.e. the fact that this was the age of the Great Discoveries with new horizons being reached in all directions. Navigators were probing routes to the Indies in both easterly and westerly directions, and Portuguese voyagers were making a thorough exploration of the African coasts, bringing back with them rich hauls of geographical information, all of which had somehow or other to be incorporated into contemporary maps. In the west, Columbus, possibly misled by the erroneous distances of Ptolemy, discovered a 'new world' instead of the anticipated Orient. All these new findings led to what in modern terms would be referred to as an information explosion. A re-thinking of Ptolemaic concepts was inevitable and compilers of world maps at the beginning of the sixteenth century endeavoured to reconcile the beliefs of Ptolemy with the mass of up-to-date data. The way in which the cartographic message was revitalised is well seen in the opening thirty years of the sixteenth century when the world map began to be revealed as we know it today.

Juan de la Cosa's world map, 1500

Although Martin Behaim of Nuremberg constructed a famous globe at the close of the fifteenth century which combined Ptolemy's views with new data gleaned from recent exploration, the outstanding world portrayal of the Renaissance was a manuscript map by Juan de la Cosa, a seaman on the first voyage of Columbus. His map was executed in portolan style on a piece of oxhide

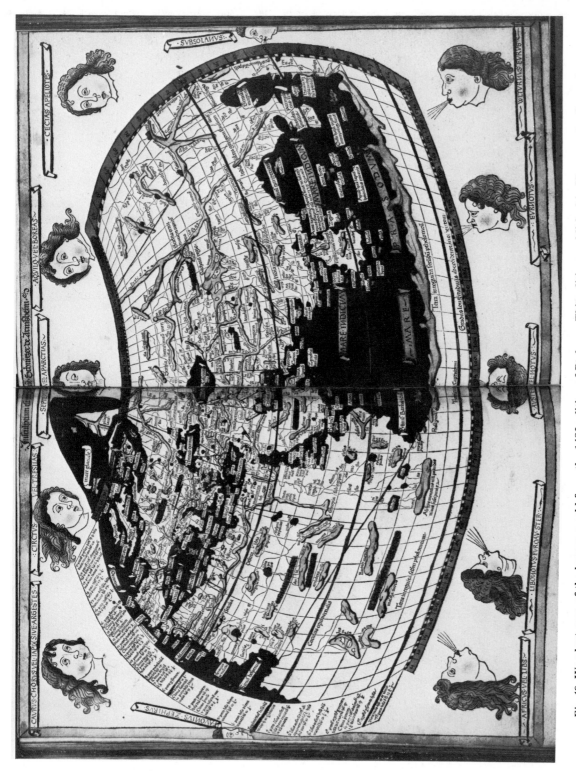

Fig 49 **Woodcut map of the known world from the 1482 edition of Ptolemy. This edition, published in Ulm, was the first to be illustrated with woodcut maps and the first to be printed in Germany. The editor, Nicolaus Germanus, follows the familiar Ptolemaic concepts; land-locked Indian Ocean, elongated Mediterranean, large island of Taprobana and no indication of the Indian peninsula. The name of the engraver, Johannes Schnitger de Armssheim, can be seen within the frame at the top of the map. This was the first map to include a credit to the engraver.**
Courtesy: **University of Liverpool   E.P.A.4.15**

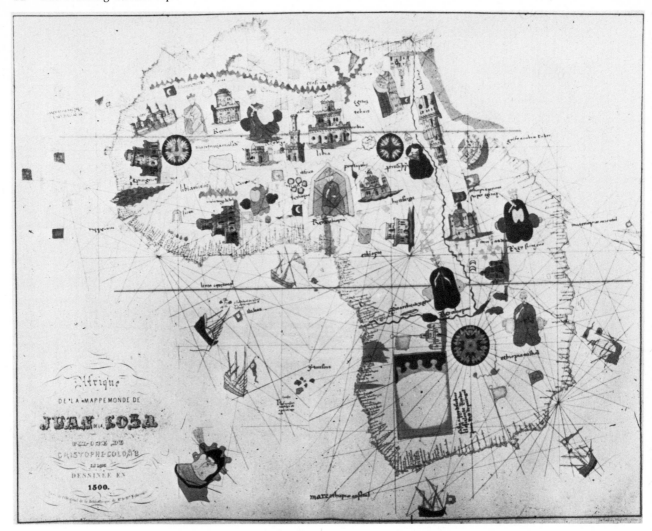

Fig 50 **Africa, a portion of Juan de la Cosa's world map of 1500, reproduced from a 19th-century facsimile by Manuel Francisco de Barros e Sousa, Viscomte de Santarem. The facsimiles of this period before the advent of photography were not faithful reproductions of original maps but hand-drawn copies reproduced by lithography.**
Courtesy: **Curators of the Bodleian Library**

three feet wide by six feet long and graphically demonstrates the initial stages in the unfolding of the American continent. It records the landfall of Cabral in Brazil, Cabot's discoveries in North America and shows the Caribbean region in some detail. Highly decorative in character, the map bears a profusion of coastal placenames with some inland detail of a pictorial nature. The Equator and Tropic of Capricorn are drawn boldly across the map and the demarcation line laid down at the Treaty of Tordesillas in 1494 to indicate the division between the Spanish and Portuguese spheres of

influence is seen running in a north-south direction. The influence of Ptolemy is seen in the poor representation of India and south-east Asia but the recent discoveries of Vasco da Gama are apparent in the improved view of Africa and the opening up of the southern flank of the Indian Ocean.

Martin Waldseemüller's map of 1507

Further improvements in the mapping of Asia were made by Giovanni Contarini in his copper-engraved world map (1506) on a conical projection. This was also the first printed map to show parts of the New World. Even greater progress came in 1507 with a superb woodcut world map by the Alsatian geographer, Martin Waldseemüller. The map consisted of twelve sheets, measuring in all eight feet by four feet six inches. It was constructed on a newly-developed projection bearing some resemblance to a projection developed by

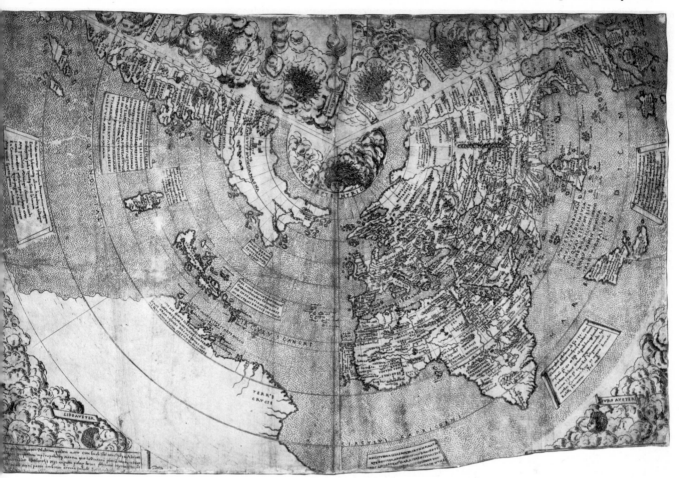

Fig 51 **A world map on a conical projection which F. Rosselli engraved for G. M. Contarini in 1506. This was the first printed map to show the new discoveries in America though they had been indicated on la Cosa's manuscript map of 1500.**
Courtesy: **British Library Board   Maps C.2.c.c.4**

Rigobert Bonne in the eighteenth century. Its content was a combination of Ptolemy-derived data and up-to-date information. One item in particular had far-reaching significance, for Waldseemüller, who had been profoundly interested in the travels of Amerigo Vespucci, placed the name AMERICA on the southern part of the continent. Waldseemüller also included a small map as an inset and it is strange that on this map he showed the Isthmus of Panama joining North and South America, while the much more detailed main map showed the two parts as separate islands. This inconsistency perhaps reveals a failure to check the finished work very thoroughly, a surprising defect in view of the high quality and ambitious nature of the map as a whole.

Diego Ribeiro's map of 1529

In 1522 the battered survivors of Magellan's attempt to circumnavigate the globe had accomplished the feat, and with their return to Seville even more radical alterations were necessary to the world picture of Ptolemy. The immense extent of the globe occupied by the Pacific Ocean was recognised and the American continent could be represented with greater reliability. Diego Ribeiro, a Portuguese who served as Royal Cosmographer to the King of Spain, produced one of the earliest maps to illustrate this new outlook on the world. In his official capacity, Ribeiro worked on the *Padron Real*, an official admiralty record of the discoveries and, although the record itself has been lost, Ribeiro's world chart provides some indication of its content. The map, as well as furnishing a cartographic record of the new geographical discoveries (and a comparison with La Cosa's map of 1500 quickly reveals the great progress made in the short space of twenty-nine years), is one of the most beautiful maps ever made. It is executed in portolan style with many names along the coasts and includes

Fig 52 **Apart from its geographical accuracy, Diego Ribeiro's world map (1529) is esteemed as a supreme example of mapmaking artistry. The illustration is reproduced at a greatly-reduced scale from a photographic copy made in 1886 of the original map which is preserved in the Grossherzogliche Statsbibliothek, Weimar.**
Courtesy: **British Library Board   Maps 184.e.3**

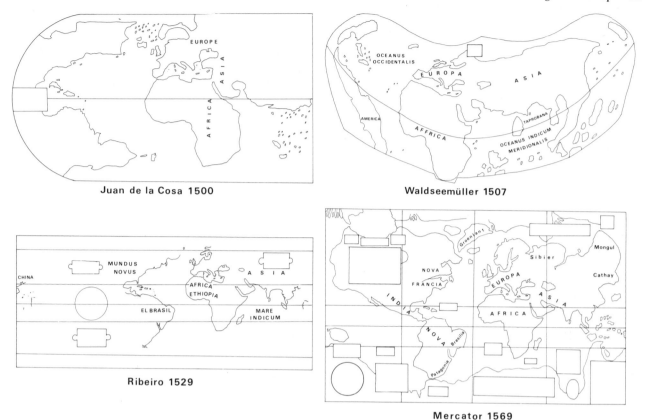

Fig 53 The world outline as seen by la Cosa, Ribeiro, Waldseemüller and Mercator.

delightfully executed illustrations of shipping and indigenous flora and fauna. In the quality of its execution and the accuracy of its cartographic portrayal, Ribeiro's map is a model of graphical communication of geographical data.

Gerardus Mercator, his projection and his world map of 1569

One of the hallmarks of the Renaissance was the 'universal genius' such as Leonardo da Vinci, who among his manifold accomplishments tried his hand at mapmaking. The Flemish cartographer, Mercator, was another such man. His diverse talents included engraving, instrument making, surveying and the manufacture of globes, as well as those of outstanding geographer and mapmaker. He is best known today for his cylindrical projection which served as the framework of his great world map of 1569. The various distortions introduced by the Mercator projection have been mentioned but on the credit side Mercator improved the delineation of south-east

Asia and Central America, made a partial correction of Ptolemy's incorrect measurement of the Mediterranean Sea, and showed particular concern about the choice of a calligraphic style which would help to convey information quickly, clearly and pleasingly. Unsatisfied with any existing script Mercator developed his own italic lettering for mapwork and wrote a treatise on it. Mercator, unlike his friend Abraham Ortelius, was an originator who extended the cartographic language and its vocabulary. Ortelius did produce a world map but is best remembered for his atlas entitled *Theatrum Orbis Terrarum* (1570) in which he assembled maps of the regions of the world from the most reliable sources available to him. *Theatrum* provided a new dimension in world mapping for it brought together in manageable form a whole series of regional maps prepared in a more or less uniform style of presentation. The regional approach enabled the world to be illustrated with far greater precision and in much more detail than was ever possible with the traditional presentation on a single map.

6 REGIONAL MAPS

Early world maps are valuable in reflecting progress in geographical thought and ideas but, apart from the concern of their makers with discovering better ways of presenting the sphere, they have rather less to impart about the development of cartographical techniques. Parallel developments in regional and local mapping convey much more information about the evolving cartographic language, particularly in the evolution of a complex system of symbols to record the multifarious facets of a restricted environment – settlements, communications, industry, land use, placenames, physiography and so on – for the greater scale at which it is possible to map regionally naturally facilitates the inclusion of very much more detail than can be done on a world scale.

THE ANCIENT WORLD

Surviving maps from Mesopotamia dating from pre-Christian times include world maps, city plans, estate plans and regional maps, the last-named category including the well-known clay tablet map found at Gar Sur, some two hundred miles north of Babylon, which has been dated at 3800 BC and illustrates northern Mesopotamia in rudimentary map form. On this small tablet, four inches in diameter, the mapmaker drew a river, presumed to be the Euphrates, flowing through a broad valley flanked by two pictorially-depicted ranges of mountains, the Anti-Lebanon to the west and the Zagros Mountains to the east. He symbolised towns by open circles and placed cuneiform inscriptions within circles to indicate north, west and east. This small regional map is one of numerous Babylonian clay tablets bearing maps which show evidence of the use of some cartographic principles in their construction, e.g. scale, orientation, symbolisation, but are really schematic diagrams or *cartograms* and bear little resemblance to a modern map.

There is little tangible evidence of other regional mapping in the ancient world but literary references indicate that maps were made of conquered territories during Egyptian military campaigns, notably that in Scythia (c. 1400 BC).

Chinese cartography has developed quite independently from that of the West and the country had been mapped in detail before it was visited by Europeans. According to Chinese tradition nine copper or bronze vases made c. 2000 BC are said to have borne representations of the provinces of the Hsia Dynasty and showed mountains, rivers and products. By 1125 BC a map of the whole country is believed to have been produced by Wen-Wang and forestry and cadastral maps were also used at this date. China was divided into thirty-six regions to replace the original nine provinces, and new

Fig 54 Clay tablet map (c.3800 BC) showing part of northern Mesopotamia with the Euphrates flowing southwards flanked by the Zagros Mountains in the east and the Anti-Lebanon in the west.

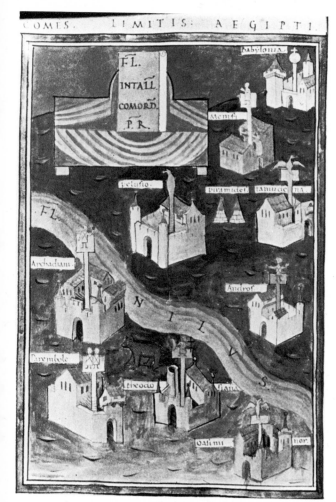

Fig 55 **Roman map from** *Notitia Dignitatum* **(5th century** AD**) showing the dispositions of garrisons in the Nile valley.** *Courtesy:* **Curators of the Bodleian Library MS.Canon. misc.378.fl13v**

The regional maps of Claudius Ptolemy

The cartographic reins in ancient Greece were firmly held by academic geographers and philosophers whose immediate concern was to establish the truth about the shape and dimensions of the earth. True regional mapping, like so many facets of cartography, began with Claudius Ptolemy (AD 87–150). Surviving manuscripts of *Geographia* are in two versions, the more usual, known as the 'A' recension, containing a world map and twenty-six regional maps, and the 'B' recension which contains sixty-four maps in addition to the world map. This regional breakdown of the world – ten maps for Europe, twelve for Asia and four for Africa in the 'A' recension – may be regarded as producing the first general atlas. It is not known which version of *Geographia* is the earlier or whether Ptolemy himself prepared any maps. In any case the Ptolemy manuscripts were lost to the scholars of Western Europe for centuries and it was only in the fifteenth century when a copy was brought to Italy by refugees from Byzantium that *Geographia* assumed its greatest significance.

Roman regional mapping

Scant evidence remains of the regional cartography of Imperial Rome. It is largely represented by the schematic route maps made to meet the needs of military and administrative authorities. Few maps of a genuinely regional character survive, one example being the rudimentary pictorial maps in *Notitia dignitatum Imperii Romani* (fifth century AD) which purport to illustrate the spatial disposition of Roman garrisons throughout the Empire.

The Madaba mosaic

In 1896 large fragments of a map of Palestine, Syria and parts of Arabia and Egypt were revealed during clearing operations on the floor of the Transjordanian church of Madaba, an important station on the Roman road linking Damascus with Petra and the Dead Sea. The surviving portions of what was originally a huge map, some fifty feet by twenty, made as a mosaic, suggest that this was a product of the time of the Byzantine Emperor Justinian (*c.*550 AD). It is an important map for several reasons. It is the oldest extant example of Christian cartography and a worthy specimen of Byzantine work which was clearly conceived as an illustration of biblical history. As such it provides a strikingly graphic picture of the Bible lands in the sixth century AD. It appears to be the first map to show the division of Palestine among the

maps had to be prepared, some of them cut in bamboo for durability, others, including those for use by travellers, were painted on silk. The earliest reference to a map made on silk relates to 227 BC. In that year the crown prince of the State of Yen induced a man named Ching Kho to attempt the assassination of the future emperor of a united China. Ching Kho, on the pretext of presenting him with a map of the district of Tu-Khang drawn on silk, also drew a dagger with which he unsuccessfully tried to kill the emperor-to-be. This story was a favourite of the decorators of the Han tombs and the earliest silk maps to survive were found in a Han tomb near Changsha in Hunan. The maps represent the area of modern Changsha and show mountains, rivers, towns and roads.

twelve tribes, the names of which were lettered prominently in red, though only six names remain. The mapmaker included considerable topographical detail
including rivers, mountains, towns and oases (for which
he used a palm-tree symbol). One hundred and thirty
placenames were included, but one of the most interesting cartographic developments is the detailed perspective portrayal of major towns, especially Jerusalem, of
which the walls, gates, churches and streets are clearly
revealed. The Madaba mosaic must be the largest map
ever made in this medium and is unique in being built
into the floor of a church.

Chinese maps on stone

As the Chinese Empire expanded over the centuries
maps at varying scales became necessary to depict the
increased territories. Among the most remarkable are
two maps, carved in stone in AD 1137. These maps are
preserved in the Forest of Steles Museum at Xian; the
first, the *Hua I Thu* (Map of China and the Barbarian
Countries) has a sketchy but recognisable coastline,
many settlements, rivers and mountains but the
unknown maker did not attempt a cartographic representation of the Barbarian lands but resorted to panels
of verbal description; the second and more sophisticated
map is the *Yü Chi Thu* (Map of the Tracks of Yü the
Great). This map is far in advance of anything produced
in Europe at its time and has a surprisingly precise
delineation of the Chinese coast and major rivers. The
map is covered by a rectangular grid with a grid scale of
100 *li* to each division. The purpose of the map was to
instruct students in the geography of the earliest survey
of China, the Yü Kung, which is roughly dated at the
sixth century BC.

The earliest known printed map was made in China *c.*
AD 1155 and pre-dates the first European printed map
by over three centuries. It was a map of west China,
printed on paper in black ink, which appeared in an
encyclopaedia the *Liu Dhing Thu* (Illustrations of
Objects mentioned in the Six Classics). This rather crude
map shows rivers and settlements and is oriented to the
north. Perhaps its most interesting cartographic feature
is a representation of the line of the Great Wall by a
double parallel line with rectangular shapes spaced at
regular intervals to indicate towers along the wall.

The climax of early Chinese regional cartography
came with a map of China prepared between 1311 and
1320 by Chu Ssu-Pén who based his work on his own
travels, literary sources and earlier maps. The map
existed for two centuries in manuscript form but was
revised in 1541 by Lo Hung-Hsien and printed in 1555
with the title *Kuang Yü Thu* (Enlarged Terrestrial
Atlas). The atlas consisted of a general map on two
pages together with numerous maps of smaller areas.
The general map, covered by a rectangular grid on a grid
scale of 400 *li* to the division, has a very evocative, and
distinctly Chinese, pattern of waves on the sea, a heavy
black band to symbolise the Gobi Desert, three parallel
lines with rectangular towers to delineate the Great
Wall, the great rivers by double lines and others by a
single line, mountains seen in elevation, and a great
number of places.

When the Jesuit fathers arrived in China during the
sixteenth century enough cartographic material was
available to produce an atlas of the Imperial lands and
thenceforward maps of China were increasingly under
the influence of European cartography.

THE MIDDLE AGES

The finest cartography of the Middle Ages in Europe
came with the highly practical, accurate marine charts
made by Genoese and Catalan chartmakers. These were
in a very real sense regional maps in their provision of a
faithful representation of the coastlines of the Mediterranean and Black Sea regions. They were, however,
essentially concerned with the sea and navigation and
are more properly discussed under the heading of navigational charts.

The regional maps which originated in the Islamic
world generally served to illustrate ninth- and tenth-
century geographical treatises, the usual format of which
included individual maps of seventeen Islamic countries,
as well as a world map, charts of the Persian Gulf and the
Mediterranean and Caspian Seas. A set of three maps of
Iran (*c.*1330) made by Hamdallāh ibn abū Bakr al-
Mustaufî al-Qazwīnī are worthy of mention in that they
consisted of rectangular grids on which no features are
symbolised though placenames are included. This
schematic grid-map treatment, or 'Mongolian style' as it
has been termed, was used in precisely the same way by
the Chinese maker of a map which appeared in the *Yuan
Ching Shih Ta Tien* (1329) or 'History of Institutions of
the Yuan Dynasty.'

The work of Matthew Paris

In Britain the religious house of St Albans, one of the
greatest monasteries of the Middle Ages, reached the
peak of its fame in the thirteenth century. Here were a
renowned library and scriptorium, and St Albans
became the major centre of English chronicles with the
monk historian, scribe and traveller, Matthew Paris, as
its most distinguished chronicler. Matthew Paris was a
gifted painter and draughtsman whose books included

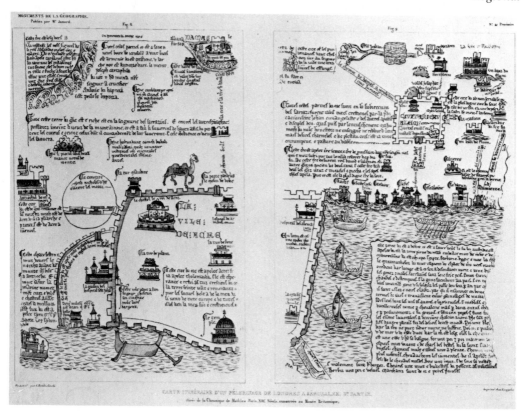

Fig 56 **A detail from Matthew Paris's depiction of the Holy Land (c.1250). The walls of the port of Acre are prominently seen, Damascus is in the top centre and Jerusalem within a rectangular wall to the right. The illustration is reproduced from a famous set of 19th-century facsimiles,** *Monuments de la Géographie***, published in Paris by E. F. Jomard.**
Courtesy: **British Library Board   Maps Ref. A3 PL.V3**

many illustrations, among them a world map, a map of Palestine and his renowned maps of Britain. The significance of the latter as route maps is separately discussed but they have also been referred to as remarkable examples of what Ptolemy called *chorography* or regional geography. Certainly for the first time there is a recognisable configuration of Britain, though a fundamental error was made in treating northern Scotland as an island linked to the remainder of Britain by a bridge. Even the best of the maps does not include much topographical detail, except for the over-prominent river system, Snowdon seen in elevation, and the Antonine and Hadrian's Walls straddling the neck of northern Britain. In general only those settlements were included which were relevant to Matthew Paris's purpose of illustrating the pilgrim route to Dover, and in keeping his map clear of irrelevant detail he showed his awareness of the fundamental principle of cartographic communication of including nothing which will distract the viewer and disturb his perception.

The Gough or Bodleian Map (*c.*1360)

Like the maps of Matthew Paris, the Gough Map can be looked upon either as a road map or simply as a regional map of Britain. In the latter context it was a great advance on any previous work and was marred chiefly by its poor delineation of Scotland. The manner in which topographical features were dealt with by the unknown maker was distinctly variable. He paid much attention to rivers, presumably because of their role in communications, but mountains, which formed a barrier rather than an aid to travel, were poorly treated. One addition to the cartographic language was the use of the pictorial symbol of intertwining trees to represent forests. Cities and towns were symbolised by groups of buildings drawn in perspective, the nature of the symbol suggesting a grading to imply the status of each town. The Gough Map is a unique example of its period and its influence was felt over the ensuing two centuries when it was certainly used as source material by Münster, by George Lily (d. 1559) in his map of 1546, and even by Mercator.

THE PTOLEMY REVIVAL

Regional mapmaking derived a considerable stimulus from the recovery of a manuscript of Ptolemy's *Geographia* in the early fifteenth century, when a copy of the

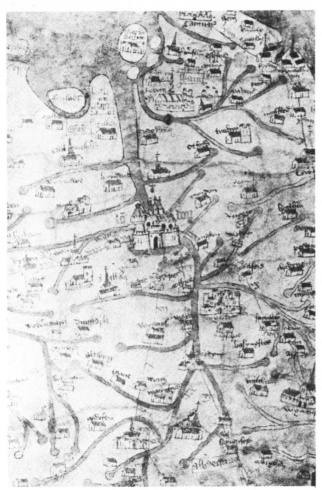

Fig 57 **Detail of the Gough or Bodleian Map (*c.*1360) showing the Thames estuary (east is at the top). London, together with several cathedral and major towns, can be seen symbolised by a group of buildings drawn in elevation.**
Courtesy: **Curators of the Bodleian Library MS.Gough Gen.top.16**

'A' recension with its twenty-seven maps was brought into Italy where it was quickly translated into Latin and the maps redrawn with Latin legends. The Latin version was recorded in several handsome manuscript copies of *Geographia* but it was only after the advent of printing, first from wood blocks and then from copper plates, that the influence of the work spread throughout Western Europe. An edition without maps was printed at Vicenza in 1475 and the first edition to appear complete with the twenty-seven maps was issued in Bologna in 1477. *Geographia* was one of the first great works, therefore, to benefit from this new invention with its far-reaching effects on the distribution of learning. It was also the first printed atlas (although the term 'atlas' to

describe a bound collection of maps was only introduced a century later by Mercator).

Ptolemy's scheme of regional maps, which was apparently the same in all copies of the 'A' recension, was a new development in cartographical studies in that he systematically divided the known world of the ancient Greeks into twenty-six regions, each of which was separately mapped. The extent of territory covered by each map, and consequently the scales, varied considerably; on the one hand Ptolemy mapped a very limited area such as the islands of Sardinia and Sicily and on the other he introduced a map covering most of South-East Asia. Of the twenty-six regional maps ten were devoted to European regions, four to North Africa and twelve to Asia.

The maps were not free from errors: for example, the configuration of the British Isles was distinctly odd with Scotland placed at right angles to the rest of mainland Britain so that it appears on an east-west axis. Ireland, too, was strangely placed, its southern coast being much further north than that of North Wales. These errors are clearly seen in early editions such as those of Bologna (1477) and Ulm (1482), the first editor to correct them being Bernardus Sylvanus who gave Scotland its correct alignment and improved the placing and size of Ireland in his Venice edition of 1511. Sylvanus was the first to amend the Ptolemaic views in the light of more recent available information. It must of course always be remembered that, despite the fifteenth- and sixteenth-century datings of early printed Ptolemy editions, the data on which they were based related to the second century AD. During the Renaissance, however, mapmakers realised the desirability of updating the Ptolemaic material so as to meet more suitably the requirements of contemporary users and to convey an up-to-date world view. They met this urgent need by introducing *Tabulae Modernae* or 'modern maps' to supplement those of Ptolemy. The first was a map of the northern regions, but principally the British Isles and Scandinavia, by a Dane, Claudius Clavus. This map was added to a manuscript edition in France in 1427. Modern maps were added to both manuscript and printed editions – the German, Henricus Martellus Germanus, added twelve new maps to his manuscripts while the printed edition with rhyming text by Berlinghieri (Florence 1482) included four modern maps. In the Strasbourg edition of 1513 the number of new maps was increased to twenty and Münster's Basle edition of 1540 included no less than forty-eight. As time went on the early Ptolemy maps became no more than a traditional adjunct to *Geographia*, a form of historical atlas which supplemented the increasing number of modern maps added to successive editions.

428

Fig 60 **Detail from G. B. Ramusio's map of Brazil in** *Delle Navigationi e Viaggi* **(Venice, 1556). West is at the top and the map, while sparse in its cartographic detail, has some interesting vignettes of native life.**
Courtesy: **University of Liverpool   Ryl.N.2.24–6**

blacked-in to the east, a technique which conveys the impression of large areas of low table-land rather than that of Alpine terrain. The outstanding French cartographer of the early sixteenth century, Oronce Finé, was deeply concerned with regional geography and the techniques of land survey. He included instructions for surveying and mapping a limited territory in a treatise, *Le sphère du monde* (1551), having already constructed a woodcut map of France in 1525.

One of the finest European regional mapmakers was the Dutchman, Jacob van Deventer, well-known for detailed town plans as well as for multi-sheet maps of Dutch provinces prepared during the 1550s and 1560s. Christiaan Sgrooten, another Dutchman with regional

interests, prepared several maps of the Netherlands and of individual provinces as well as a map of Palestine (a country which received singular attention from mapmakers on account of its unique role in history).

THE ATLASES OF ORTELIUS AND MERCATOR

The volume of geographical data emerging during the latter half of the sixteenth century necessitated a new means of accommodating this accumulated knowledge which would be more manageable than the existing series of sheet maps. The answer was the atlas, a bound collection of uniformly-sized maps arranged in systematic order. The Flemish geographers, Abraham Ortelius (1527–98) and Gerardus Mercator (1512–94), saw the advantages of producing world atlases which would gather together in a relatively handy format the summation of regional cartography. The atlas concept was not, of course, entirely unknown, for the printed editions of Ptolemy which included maps were in essence world atlases on a regional basis; so too were the miscellaneous collections of maps assembled by sixteenth-century Italian mapsellers such as Lafreri to meet individual customers' requirements. These *Lafreri Atlases*, as they came to be known, had two disadvantages; first, the format could vary from map to map – some large maps, some small – and secondly, maps of the same region by several different authors might appear in the same collection. Ortelius insisted on complete uniformity of format and used maps by one author only for each of his regions. His atlas, a major landmark in the graphical communication of geographical data, was published in 1570 with the title *Theatrum Orbis Terrarum*. Its seventy maps were collected by Ortelius from the most reliable sources – in other words, he was a scholarly editor rather than an original mapmaker and he acknowledged this by listing the names of those authors whose work he used. *Theatrum* ran into many editions in several languages, some of the later editions being considerably extended; the 1601 edition, for example, included one hundred and twenty-one maps compared with the original edition's seventy.

Mercator's *Atlas* was even more ambitiously planned, so much so that its creator did not live to see the publication of the entire work. Mercator began to publish the atlas in parts; Gallia-Belgia-Germanica in 1585 and Italia-Slavonica-Grecia in 1589 – but the entire set of one hundred and seven maps was not issued until 1595, a year after Mercator's death. From this date onwards, for a century and more, the atlas became the dominant cartographic form and the seventeenth century passed into mapmaking history as the 'age of the atlas'.

THE DUTCH PUBLISHING HOUSES

The middle and later periods of the seventeenth century were dominated by the Dutch publishing houses of Blaeu, Jansson and their successors, Visscher, De Wit, Schenk and Valck. Their elegant atlases, wall and sheet maps are sometimes dismissed as making little contribution to cartographic progress, and though this may be true as far as content is concerned, there can be no doubt that the effectiveness of the map as a communications medium was enhanced by the clear, logical layouts, the fine engraving and superb calligraphy of seventeenth-century Dutch mapmakers. The technical side of making and reproducing maps was also improved; Blaeu, for instance, attained his personal ambition of housing under one roof an establishment in which every phase of map production could be carried out efficiently. To this end his printing house contained presses for letterpress printing of text matter, presses for copperplate printing and a type foundry.

The general Dutch approach to cartography, therefore, was to concentrate on improved quality with the emphasis particularly on artistry and a high standard of craftsmanship.

CHRISTOPHER SAXTON AND THE DEVELOPMENT OF THE BRITISH COUNTY MAP

The late sixteenth and early seventeenth century was a period of radical change in the mapping of Britain. British mapmakers, whose contributions to cartographic progress had been rather meagre, took the lead in a new approach to regional cartography. Previously the country had been mapped as a single unit with excellent maps by Lily, Mercator, Ortelius and others. In 1563, however, Laurence Nowell, in a letter to Sir William Cecil, complained of the inaccuracy of the general maps of England and outlined his intention of constructing maps of the separate counties. Two sets of manuscript maps by Nowell do exist but it was a Yorkshire estate surveyor,

Christopher Saxton, who pioneered that most distinctive British cartographic form, the atlas of county maps. Saxton's contribution to the regional cartography of Britain is immense and his *Atlas of England and Wales* (1579) was not only the first atlas of the country but the first *national* atlas produced anywhere. Blessed with the patronage of the wealthy Thomas Seckford and the support of Queen Elizabeth, Saxton rapidly surveyed and mapped every county, unhindered by any of the financial worries that thwarted some of his contemporaries. His maps, printed from copper plates engraved by Flemish and English craftsmen, set a standard for the next two centuries and their content influenced series upon series of later county maps. In style and ornamentation the general map and thirty-four maps of Saxton's *Atlas* followed the Flemish pattern and in some instances – the map of Cornwall, for example, engraved by the Fleming, Terwoort – the ornamentation is so profuse that the attention is diverted from the map itself with a consequent lessening of effective communication.

John Norden, an equally talented surveyor, devised his own scheme for a national atlas but failed for lack of patronage, only five maps being published. Nevertheless he was responsible for such innovations as boundaries of hundreds, a gazetteer of placenames, a reference grid and a key to the symbols he used. For the last named feature he no doubt derived inspiration from the herald/topographer, William Smith, who had lived in Nuremberg and was influenced by German traditions. Smith produced an impressive series of twelve county maps in 1602–3 which had the advantage of superb engraving by Jodocus Hondius of Amsterdam, a master craftsman who added much to the visual appeal of the county maps in *Theatre of the Empire of Great Britaine* (1611–12) by John Speed. Apart from the improved composition, calligraphy and engraving compared with Saxton's maps, those of Speed featured only one real innovation, the inclusion of a town plan as an inset on each map. Speed was not an original surveyor but a compiler or editor in the mould of Ortelius and he freely acknowledged his indebtedness to other men's work.

The work of these pioneers of British regional cartography meant that the country was mapped in unprecedented detail but, surprisingly, instead of initiating an era of steady progress their efforts became a source of plagiarism with ensuing mapmakers content to derive their data from the early works, including all the errors and inaccuracies. In the mid-seventeeth century the work of Blaeu and Jansson introduced a superior quality of presentation to British county mapmaking and another step forward came with the publication of John Ogilby's roadbook *Britannia* in 1675, after which roads became an important feature of county maps.

Fig 61 **Superbly engraved map of Calabria from the 1606 edition of Ortelius's *Theatrum Orbis Terrarum*. The map shows the elegance caused by the introduction of copper engraving, the craftsman having achieved a perfectly even texture by his stippling of the sea areas and his calligraphy is of the highest standard, including flourishing 'swash' letters for the sea names.**
Courtesy: **University of Liverpool   Ryl.N.1.3**

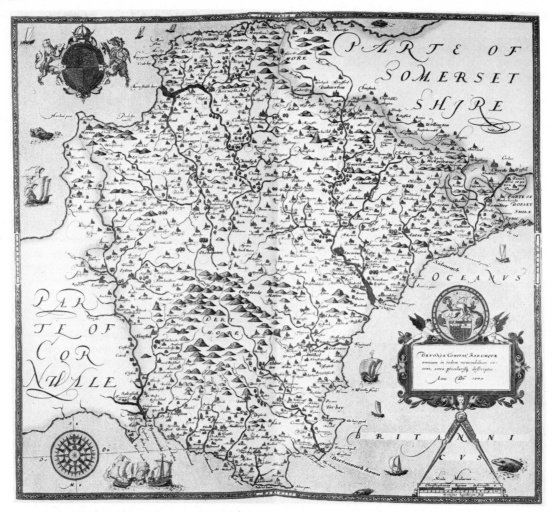

Fig 62 **The county of Devonshire from Christopher Saxton's** *Atlas of England and Wales* **(1579) engraved by the Flemish craftsman, Remigius Hogenberg.**
Courtesy: **British Library Board   Maps C7.c.1**

As far as British cartographic development was concerned there was a marked contrast between the two halves of the eighteenth century. There was relatively little originality before 1750 and much reworking of earlier material. The second half of the century, however, began propitiously when the Society for the Encouragements of Arts, Manufactures and Commerce (known since 1847 as the Royal Society of Arts) offered an award of not more than £100 for an original county survey at the one inch to one mile scale. The offer stimulated the preparation of fine county maps which surpassed in accuracy and wealth of detail the maps which had so far appeared in atlases of county maps. The first winner was Benjamin Donn's map of Devon (1765) based on an accurate survey costing almost £2000. In all, prizes were awarded for thirteen county surveys and the

Society deserves credit for improving the quality of British county mapmaking and to a great extent for bringing about a standardisation of format. The Society's examination of the maps submitted to them must have led to greater accuracy in survey and certain maps bear witness to their scientific construction by the inclusion of a survey triangulation diagram.

With the inauguration of the Ordnance Survey in 1791 the writing seemed to be on the wall for the private regional surveyor, but the making of county maps continued unabated for the first half of the nineteenth century, before gradually dwindling away up to 1900. At the turn of the nineteenth century men such as John Cary and Charles Smith produced small, reasonably priced atlases which were very popular, and private surveyors such as Christopher Greenwood and Andrew Bryant had their own schemes for one-inch scale surveys of the whole nation. Commercial cartography as carried out by Greenwood and Bryant depended on securing subscribers in advance by means of advertisements in local newspapers, followed by a distribution of prospectuses

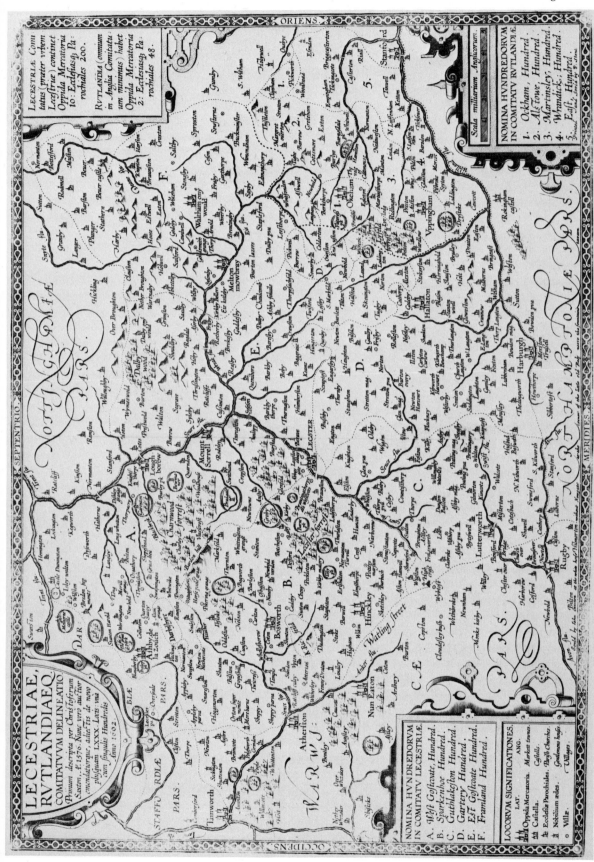

Fig 63 County map of Leicester and Rutland by William Smith (1602–3). This was one of twelve maps now attributed to Smith which were unidentified until recently and known as the 'anonymous series'. Smith, unlike Saxton and Speed, provided a key to the symbols used on his map. The illustration is from a late reprint by Overton and Stent.
Courtesy: University of Leicester

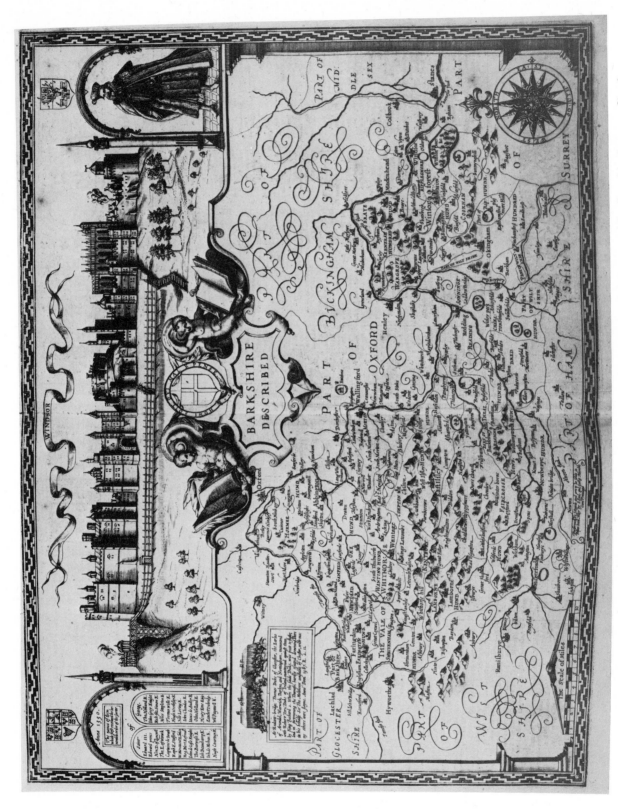

Fig 64 Berkshire from John Speed's *Theatre of the Empire of Great Britaine* (1611). This is unusual for Speed in that a profile of Windsor Castle is substituted for the customary inset town plan. The Berkshire town of Reading is in fact featured on the map of the adjacent county of Buckinghamshire.

Bodleian University of Liverpool. H.17.2.

Fig 65 **Detail from Benjamin Donn's prize-winning map of Devonshire (1765). Prepared at the one-inch scale, the map includes much topographical detail and features names of prominent landowners who would be expected to subscribe to the map.**
Courtesy: **University of Exeter**

to major landowners in each county. The necessity of procuring the patronage of county gentry exerted some influence on the content of the county maps whose makers paid particular attention to the inclusion of the houses, surrounding parks and interests of the landowners. In their attempts to provide a faithful delineation of the face of a county the early-nineteenth-century county mapmakers were mainly deficient in their portrayal of relief which was ineffectively displayed by hachuring. In other respects the maps of Greenwood and Bryant are a useful source of information about boundaries, land use, the impact of industry and the communications system.

THE CARTOGRAPHY OF NORTH AMERICA DURING THE EIGHTEENTH AND NINETEENTH CENTURIES

European mapmakers found a fruitful field of operations in North America during the eighteenth century when the organisation of the colonies, the growth of settlement and Anglo-French rivalry led to a demand for reliable regional maps. The Frenchman Guillaume de l'Isle and his successors Buache, N. H. de l'Isle and J. A. Dezauche all made contributions to the mapping of the Americas; Guillaume correctly calculated the longitudes of America, corrected the current fallacy of California as an island, properly delineated the Mississippi valley and introduced many new placenames; Buache was the first to suggest that America and Asia had once been joined at the Bering Straits; Nicholas de l'Isle and Buache delineated a fictitious coastline for north-western America. The Dutchman, Hermann Moll, the Englishman, John Senex, and the Germans, Homann and Seutter, all included maps of parts of North

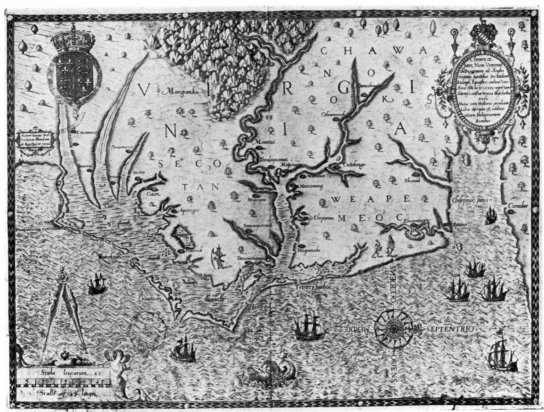

Fig 66 Map of Virginia drawn by John White and engraved by the Frankfurt printer and engraver, T. de Bry, for *America,* **Part I (1590). This map was the basis of European maps of the area for over 80 years. White made several voyages to America and became governor of Roanoke Colony in 1587. His map first appeared in Thomas Harriot's** *A briefe and true report of the new found land of Virginia* **(1588).**
Courtesy: **Curators of the Bodleian Library   J.Maps 230 (1A)**

and South America in their general atlases but it was in 1733 that Henry Popple issued a really detailed map of North America in twenty sheets. The peak of colonial mapmaking is regarded as being Dr John Mitchell's *Map of the British and French Dominions in North America* (1755), a very large, detailed map which has its own particular place in American history, for it was used by the Peace Conference of Paris in 1783 when the boundaries of the United States were laid down on it. Mitchell was a distinguished botanist and physician who emigrated from England to settle in Virginia *c.*1700. His map was engraved in London by Thomas Kitchin and is especially noteworthy for its many inscriptions which not only attempt to justify British claims to regions in dispute with the French but also provide interesting information about places, falls, rivers, Indians and so on. An inscription south of Lake Erie reads, 'The Places

called Licks, and Lick Creeks, are Salt water, which afford plenty of Salt to Man & Beast in those Inland Parts, The Resort of all sorts of Game, Huntsmen, Traders & Warriors, especially the Salt Ponds', and one west of the Missouri informs that 'The Padoucas and Panis are reckoned very numerous, the Panis are said to have 60 or 70 Villages'.

During the Revolutionary War a large number of military maps were produced to demonstrate troop movements, plans of forts, battle strategy and the course of campaigns over large areas. Several important regional maps were prepared, sometimes based on those of twenty years earlier, but adding topographical information such as roads which would be of value in military planning.

British mapmakers in North America during the seventeenth and eighteenth centuries left a considerable legacy for their American successors. British mapmaking skills had set a standard for others to follow and British maps played an important part in disseminating knowledge of the North American territories.

At the close of the Revolutionary War in 1783 the young Republic faced the task of subdividing and settling the vast tracts of land it had inherited. A Public Domain was created in the west for the mutual benefit of the thirteen former British colonies along the Atlantic Coast and during the seventy-five years after the war this

Fig 67 Detail from a facsimile of John Mitchell's *Map of the British and French Dominions in North America* **(1755). This map which exercised a considerable influence on later maps is representative of its period in concentrating on map detail and eschewing extraneous ornamentation.**
Courtesy: **Harry Margary and the University of Cambridge**

was gradually expanded to encompass the entire territory from Atlantic to Pacific. Both George Washington and Thomas Jefferson were keenly interested in survey, and Jefferson's *A Map of the Country between Albermarle Sound and Lake Erie* (1787) was an important example of American map compilation of its period. Jefferson was chairman of a committee established to organise the settlement and subdivision of the new territories, and the Land Ordinance of 1785 was produced as a result of this committee's report. Thomas Hutchins, a former army surveyor, was appointed Geographer of the United States with the appointed task of carrying out a cadastral survey and he began the United States Public Land Survey in eastern Ohio. The system adopted was to divide land into mile-square sections, the blocks being controlled by principal meridians running north-south, and base lines running east-west. Progress was slow at first but as settlers demanded more land, the clamour for

maps increased and by the time of the American Civil War (1861–5), considerable territory had been surveyed and the General Land Office was one of the most active official mapmaking agencies. This great land survey was the most remarkable example of one of the most significant trends in nineteenth-century mapmaking, the cadastral surveying of vast areas of territory.

After the Civil War, maps were needed for a variety of legal and administrative purposes and the result was a prolific output of cartographic products, in which county atlases, based on Land Office Surveys supplemented by original information assembled by the publishers, were to the forefront. The typical American county atlas was sold by subscription and featured detailed township maps with the boundaries of all properties and details of ownership clearly indicated. County atlases were still produced at the beginning of the twentieth century but rapid urbanisation and the expansion of industry caused them to lose much of their former significance. Nevertheless, they provide modern researchers with a valuable source of information about the agricultural society of their period.

The cartography of Canada developed along two distinct lines during the latter part of the nineteenth century. In the populous eastern regions privately published county atlases and county maps flourished. Elsewhere

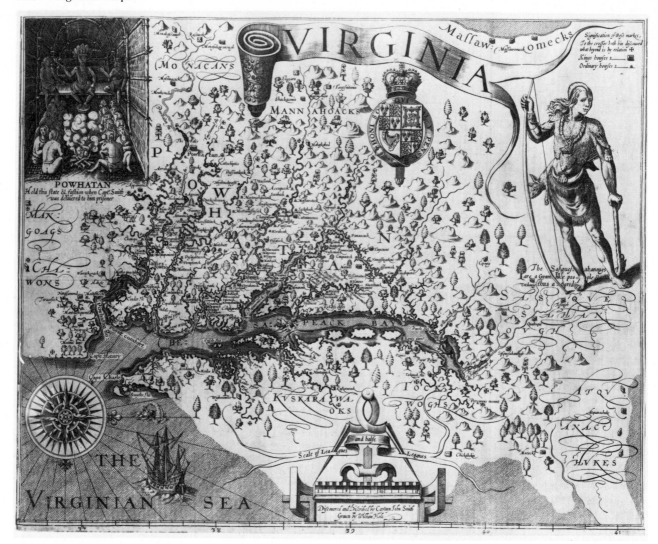

Fig 68 **Captain John Smith's renowned map of Virginia was engraved by the Englishman, William Hole, in 1612. It was the first map to provide a reasonably accurate view of Chesapeake Bay. Hole's engraving of Powhatan is based on de Bry's engraving of a drawing by John White.**
Courtesy: **British Library Board   G.7121**

official mapmaking was predominant with the Geological Survey making its own topographical maps from 1842 and the Dominion Lands Survey commencing work in 1867 on a detailed map coverage of the west prior to the building of a railway across the continent and the movement of settlers to the Pacific coast areas.

REGIONAL MAPPING IN THE MODERN WORLD

At the present, world mapping and survey is normally carried out at national level by official agencies which produce maps on a systematic arrangement of sheet lines. Survey is based on a national triangulation framework to provide a continuous cover of an entire nation. In this respect contemporary regional mapping differs from that of the early private surveyor who produced a series of county or provincial maps, each of which was a quite separate entity, possibly differing in scale from the maps of neighbouring territories and supplying virtually no detail between the regional boundary and the frame of the map.

The United States Commission on National Atlases has done a great deal to encourage the preparation of atlases, at both national and regional level, which synthesise the geography of the relevant territory. Recently-produced European atlases include the outstanding ten-volume *Deutscher Planungsatlas*, commenced in the 1950s. Outside Europe, provincial atlases of exceptional quality have been published. The *Economic Atlas of Ontario*, for example, has been widely acclaimed, not only for its cartographic expertise, but also for its beauty.

7 NAUTICAL CHARTS

From earliest times men have recognised the need for landmarks and erected some form of construction which would be instantly recognisable and allow them to establish their whereabouts. When travelling overland man made piles or cairns of stones to mark his route and if he ventured out to sea bonfires would be erected along the coast to assist his safe return to harbour. Primitive lighthouses were a natural development, the earliest being built on the island of Pharos near Alexandria about 250 BC. This great tower, some six hundred feet high, was regarded as one of the seven wonders of the world and appeared prominently on medieval maps such as those of Hereford and Ebstorf. Apart from such lighted landmarks, early mariners were almost entirely lacking in navigational aids and groped a hesitant way around the coast using log and lead-line. When navigating close to the coast seamen had to know well the salient coastal features, the distance between headlands and bays and the position of any hazards. In earliest times such information was stored in the seaman's head and passed on to others by word of mouth.

Coastal voyages, then, could be undertaken with reasonable confidence, but if the mariner ventured further afield into open waters a degree of skilled navigation was necessary. Four thousand years ago people from Peru migrated across the Pacific Ocean using rafts of balsa wood, and Polynesian and Micronesian peoples voyage from island to island in their catamarans. Before embarking on any such adventure it was essential that the mariners should know in which direction their destination lay and that when out of sight of land they should be able to maintain a correct course in this direction. Early seamen studied the movement of sun and stars and the direction of prevailing winds for this purpose. The Polynesian and Micronesian navigators were particularly skilled in reading the wave and eddy patterns created by islands hundreds of miles distant. It was often important to know when a landfall could be anticipated and the Polynesians carried pigs on their vessels because the animals became excited when close to land and

sniffed in the direction of the scent. Another stratagem was to carry land birds which could be released to see in which direction they would fly. In the Mediterranean region seamen placed great reliance on eight prevailing winds which maintained remarkably constant direction, and it is not surprising that directions in the area became known by the names of these winds and were so depicted on maps.

In classical times Mediterranean seamen had the use of written information provided in a book called a *periplus* or coast pilot, the earliest extant example dating from *c.*450 BC. The first *periplii* were accounts of voyages, with detailed descriptions of coastlines and harbours and information about distances between ports. Later works provided information about tides, wind directions, sandbanks, rocks and other hazards. The availability of such information in an accurate form could make the difference between a safe landfall and being lost at sea. Safe navigation was truly a life and death matter.

One important step forward in rendering the mariner's life less hazardous was the introduction of the mariner's compass into the Mediterranean region, probably during the eleventh century. This instrument freed him from his dependence on clear skies and made him reasonably confident of safely reaching his destination. Originally conceived as a sliver of magnetic ore stuck into a reed and floating in a bowl of water, the compass was later developed into a needle rotating on a brass pin with directions marked around the rim of the bowl. With the aid of the new compass a helmsman could keep a constant check on his course, and by the fifteenth century this had become the established method of maintaining direction when out of sight of land.

PORTOLANO

The available aids now enabled the medieval seamen to establish direction, maintain a course, calculate speed of travel and gauge the depth of water. So far no attempt had been made to assist the mariner cartographically, but the old periplus began to assume a more sophisticated form, a navigational guide called a *portolano* which provided detailed written information about coastlines, harbours and offshore hazards. These portolani began as private notebooks kept by mariners, but when passed on to others they constituted an accumulation of invaluable practical knowledge and experience. The *rutters* of later English sailors, the *routiers* of the French and the *leescaert* of the Dutch were equivalents, though not necessarily descendants, of the Italian portolani.

PORTOLAN CHARTS

At the close of the thirteenth century the portolani were supplemented by charts which laid down the information supplied by the written directions in a cartographical format. The new manuscript charts, known as *portolan charts*, were practical aids designed to communicate vital information to seamen about coasts and harbours in a highly graphic manner. There are, of course, certain differences between the making and the usage of a nautical chart and that of a topographical map; the chart is essentially a route-finding device which describes everything a mariner should know about a particular area of sea – water depths, shipping hazards, compass variations, currents, buoys, lights and so on. New symbols are required to indicate these features. The gathering of the basic information from which the chart is to be constructed is difficult, for the lack of landmarks when out at sea means that reliance has to be placed on celestial observations and, as the centuries went by, Arabian and European astronomers invented more sophisticated, accurate instruments for taking the altitude of celestial bodies. Of these, the sextant – independently invented in England and the United States in the eighteenth century – remains in use today; it is basically an instrument for measuring the vertical angle of heavenly bodies from the horizon and is used, in conjunction with a chronometer, navigational tables and the Nautical Almanac, to determine a ship's position in terms of latitude. Another important way in which it is desirable that a nautical chart should differ from most land maps is in the choice of projection, for the mariner needs a chart in which all straight lines are of constant bearing. Such a projection was provided by Mercator in 1569.

The new portolan charts were the work of artist/craftsmen and based on firsthand experience, combined with written and verbal reports, rather than on scientific observations. Usually drawn on a piece of vellum, with the neck to the left, they were often beautifully executed and coloured. Their construction was based on the known direction and distance from one harbour or coastal feature to the next. Coastlines were drawn in freehand with surprising accuracy but the most striking visual feature of most portolan charts is the network of *rhumb* or direction lines which radiate over the whole area from a system of strategically-placed compass roses, normally with two centrally-located major roses and interconnecting subsidiary roses arranged as an octagon around the periphery. These rhumb lines, or lines of constant bearing, were superimposed to aid the navigator to lay his course using dividers and straightedge.

Chart production was concentrated on two areas, the north Italian ports of Genoa and Venice and the Spanish ports of Palma and Barcelona. One of the main differences between the Italian and Catalan charts lay in their range; Italian charts usually portrayed Western Europe with the Mediterranean and Black Seas but the Catalan charts included Northern Europe and Scandinavia. The well-known coasts of the Mediterranean and Black Seas were delineated with great accuracy, those of the north much less so. A wealth of coastal names was always included, drawn inland from the coastline and perpendicular to it. Names of ports were in red, other names in black, while particularly important places were sometimes distinguished by a banner or portrayed in bird's eye view. The coast itself was quite faintly shown and stood out chiefly because of the multitude of names crowded along it. Islands appeared prominently with brilliant colouring in reds, blues, purple and gold. River deltas were shown as a series of islands and provided a colourful display. It was important, of course, and still is today, that however much information is provided on a marine chart it must be immediately legible and also essential, for the inclusion of extraneous matter may mean that the navigator is unable to plot his course clearly. The portolan charts, though fulfilling a number of the navigator's requirements, were not drawn on any projection, their makers ignoring the earth's sphericity and providing no indication of parallels or meridians. A linear scale of distance was often provided, however.

The earliest known surviving chart is the so-called *Carte Pisane* or Pisan chart dating from *c*.1300, a plain work in appearance with no compass roses. The earliest chart to bear a date is that of Petrus Vesconte, made in Florence in 1311, but the finest fourteenth-century chart is that made by Angellino de Dalorto (Florence 1325), a chart which includes more than the usual modicum of inland topography and is finely drafted and coloured.

THE CATALAN ATLAS OF 1375

The zenith of Catalan cartography was reached with the *Catalan Atlas*, a work which is regarded as one of the

the plates occupied twelve years and that five thousand pounds of copper were needed.

DE SPIEGHEL DER ZEEVAERT

The first printed sea-atlas was run off on the Leiden press of the famous Antwerp printer, Christopher Plantin, in 1584. *De Spieghel der Zeevaert* by Lucas Janszoon Waghenaer, an Enkhuisen seaman and pilot who later in life became a customs officer, is one of the landmarks of marine cartography. Designed for pilotage rather than general navigation, the atlas appeared in two parts containing forty-four charts taking in the coasts of Northern and Western Europe. It was highly popular with mariners, so much so that it ran into many editions and in England it was so influential that all sea-charts thenceforward were known as 'waggoners'. The highly decorative charts were engraved on copper by Baptist and Johannes van Deutecum and printed with meticulous care by Plantin. Waghenaer conveyed useful information to the seaman by means of a coastal profile drawn as it would be seen from a ship running along the coast. On the charts themselves he illustrated coastal towns in the fashion of the day by pictorial groups of buildings and inland he attempted a panoramic view of the surroundings, shading hills on their right-hand slopes, drawing fields in perspective and symbolising woodland by clusters of trees. Offshore he provided numerous soundings and showed anchorages (by a small anchor symbol), dangerous rocks and buoys. Each chart was lavishly ornamented with elaborate cartouches, sea monsters, sailing vessels and so on, to such an extent as to produce 'cartographic noise' which may have distracted the mariner's attention from more important factual matters.

An anglicised version of the atlas appeared in England in 1588 under the title, *The Mariners' Mirrour*. The translation was made by Anthony Ashley, Clerk to the Privy Council, and the Dutch charts were re-engraved with English cartouches and lettering by four eminent craftsmen – De Bry, Hondius, Rutlinger and Ryther. Published at a particularly apposite time in view of the threat from the Armada, the atlas furnished information about British coasts which had hitherto been unavailable to British seamen. It seemed, therefore, that the Dutch were better informed about the hazards of coastal navigation in British waters than were the natives.

THE SEVENTEENTH-CENTURY DUTCH MONOPOLY

The early seventeenth century saw Waghenaer meeting with intense competition from Dutch rivals beginning with *Het Licht de Zeevaert* (1608) by the leading Dutch publisher, Willem Janszoon Blaeu. Blaeu introduced some improvements: his coastlines were more accurate and Waghenaer's tendency to present the more important coastal features at an exaggerated scale was discarded; he included many more lights, beacons and buoys (an indication of the improvement in navigational aids between 1584 and 1608). Blaeu also removed the coastal profiles which were such a useful feature of Waghenaer's charts and inserted them within the associated text.

His most formidable rival, Jan Jansson, published sea-atlases from 1620 onwards in direct competition with Blaeu and in many cases his charts were close copies. Volume v of his *Atlas Major* was a sea-atlas. Many publishers entered the lucrative field of chartmaking from 1650 onwards. *De Lichtende Columne* (1650) by the Amsterdam engraver, Pieter Goos, is one of the most beautiful sea-atlases of this period, Goos's charts, like those of his competitors, being hand-coloured whenever a customer so desired and was willing to pay the extra cost. Other Dutch hydrographers active during the late seventeenth century included Jacob Aertz Colom and his son Arnold, Hendrick Doncker and Johannes van Keulen.

The Dutch monopoly remained unchallenged until John Seller published *The English Pilot* (1671–2), but this work was heavily criticised for its inaccuracies, not surprisingly, as Seller relied heavily on the Dutch charts for his information.

'GREAT BRITAIN'S COASTING PILOT' (1693)

Samuel Pepys, Secretary of the Navy, complained that Seller 'had bought the old worn Dutch copper plates for old copper and refreshed them in many places'. Clearly an entirely new survey of British coasts was urgently required and Pepys issued an Admiralty Order in 1681 commissioning Captain Greenvile Collins, an experienced naval officer, to survey the coasts and harbours of Britain. Collins achieved his enormous task in the surprisingly short space of seven years, using comparatively crude survey techniques – measuring chain, compass and lead-line – in a survey which could not be related to any national triangulation for the simple reason that none existed in Britain at that time. Forty-eight charts finally appeared in Collins's *Great Britain's Coasting Pilot*, a work which achieved fourteen further editions during the eighteenth century. The charts were plane charts and not prepared on Mercator's projection. Though superior to Seller's charts, those of Collins were criticised by Pepys who wanted members of Trinity House to review all Collins's work with a view to correcting its

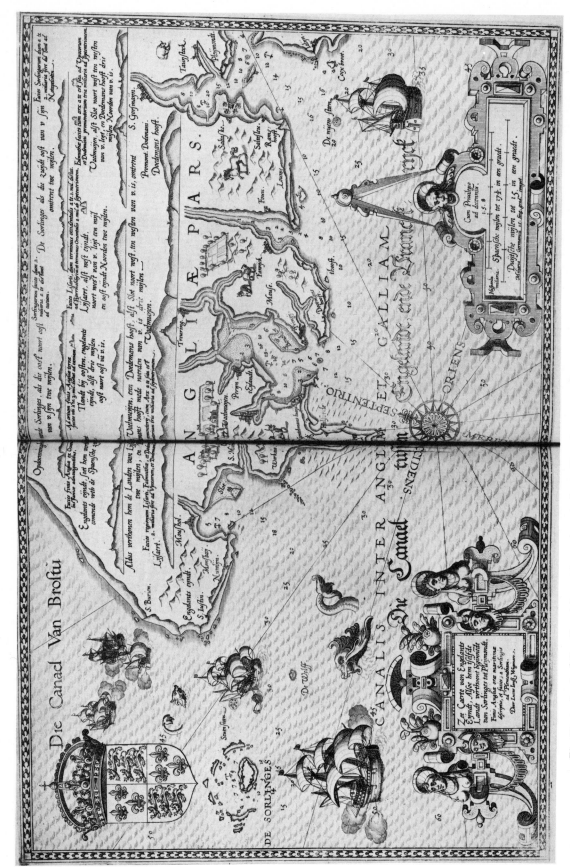

Fig 72 A chart of the Cornish coast, westwards from Plymouth, from L. J. Waghenaer's *De Spieghel der Zeevaert* (1584). Beautifully engraved with delightful pictorial information and a series of coastal profiles to assist the mariner in identifying his position.
Courtesy: **British Library Board   Maps C8.b2**

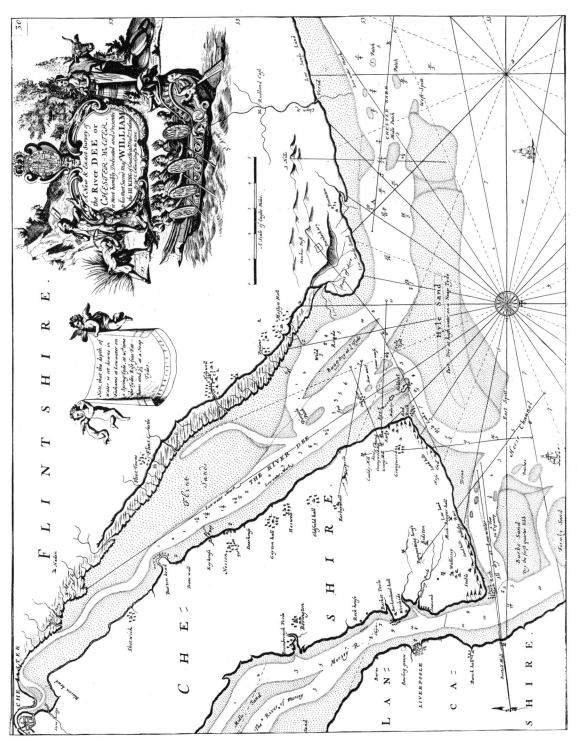

Fig 73 **Chart of the Dee estuary from Captain Greenvile Collins's *Great Britain's Coasting Pilot* (1693). An interesting feature is the series of straight lines originating in small circles and ending in prominent landmarks on shore. The circles represent the position of a survey vessel, the locations of which were fixed by compass bearings represented by the straight lines. Soundings were taken by lead line and depths indicated in fathoms.**
Courtesy: **University of Liverpool**

inaccuracies. Nevertheless, despite its faults, Collins's atlas is a landmark in British hydrography, being the first pilot book produced with English text to cover the entire British coastline.

'LE NEPTUNE FRANÇOIS'

In the late seventeenth century the French, under Louis XIV, took the initiative in a new scientific approach to cartography. The *Académie Royale des Sciences* was set up in 1666 expressly to relate astronomical observations to the problems of geography, cartography and navigation. In Paris the Italian cosmographer, Jean Dominique Cassini, decided on the Galilean method of observing the movement of Jupiter's satellites as the best method of determining longitude on land. He also gave attention to the method termed 'transport of chronometers' in which longitude could be calculated by comparing local time with Paris meridian time on the basis that 15° of longitude is equivalent to one hour's difference in time. Cassini, whose sons and grandsons also made major contributions to the mapping of France, was further responsible for a precise survey of France, using the triangulation method described by Frisius in 1533. The combined achievements of the Cassinis were to have far-reaching effects, both on French cartography and further afield.

In 1693 Charles Péne was authorised by Louis XIV to publish an atlas of twenty-nine charts entitled *Le Neptune François*, the charts being based on official French surveys and drawn on Mercator's projection. In the same year the eminent French geographer, Alexis Hubert Jaillot, collaborated with the Dutch publisher, Pierre Mortier, to print an edition in Amsterdam, though Paris was given as the place of publication on the title page. The most important aspect of this new atlas was the precision with which coastlines were delineated as a result of tying-in coastal features to the accurate Cassini land survey. Indeed it is reported that Louis XIV was piqued at the way in which the boundaries of France seemed to have shrunk as a result of the new accuracy. In 1751 the atlas plates were acquired by the French Navy Department who commissioned an official revision. This appeared in 1753 under the editorship of J. N. Bellin, a fine cartographer best known for his *Petit Atlas Maritime* (1764), a five-volume work with 581 charts.

HARRISON'S MARINE CHRONOMETER

Britain, though generally overshadowed by the French, made some contribution to the eighteenth-century scientific approach to cartography and navigation. In the early years of the century the problem of determining longitude at sea seemed as far from solution as ever, and in 1714 the British Parliament passed the Longitude Act offering £20,000 to anyone who could find the answer. As far back as 1522 Gemma Frisius, a thinker well ahead of his time, had indicated that an accurate timepiece was essential for navigation. The pendulum clock was unusable in an unsteady ship and a Yorkshireman, John Harrison, devoted thirty years to developing a marine clock which would be controlled by a spring escapement rather than by a swinging pendulum. In 1763, at the fourth attempt, he produced a timepiece which during 156 days of sea trials lost only fifteen seconds. The longitude problem could now be solved, and the way cleared for the construction of accurate, larger-scale charts which would make the navigator's task easier and safer.

THE 'AMATEUR' HYDROGRAPHER IN EIGHTEENTH-CENTURY BRITAIN

Britain, a late starter, began to take her place in the story of the evolving marine chart during the eighteenth century, and an interesting feature of British hydrography was the activity of numerous private individuals who combined an aptitude for mathematics with a flair for survey. Prominent among this group was the Holyhead customs officer, Lewis Morris, whose ambitious plan to survey the west coast of Britain was only partially realised due to opposition from Liverpool shipowners who were committed to support a survey of the Lancashire and North Wales coasts by Samuel Fearon and John Eyes. Morris, however, did obtain Admiralty agreement for a survey of the Welsh coast from Llandudno to Milford Haven and his *Plans of harbours, bays and roads in St. George's and the British Channel* (1748) included trading statistics, suggested modifications to Welsh ports and a Welsh vocabulary for use by traders, to supplement charts which in themselves were an improvement on anything previously available.

MURDOCH MACKENZIE

The surveys of Murdoch Mackenzie, born in 1712 in Orkney, were a notable advance in British hydrography. In 1744 Mackenzie began to make the first truly accurate survey of British coasts, a survey based on a precisely determined base line on land and a network of triangulation, the angles of which were measured with a theodolite sighted on to beacons specially erected on hilltops. The results of these surveys were published in 1750 in *Orcades, or a Geographic and Hydrographic Survey of the Orkney and Lewis Islands*. Mackenzie was commissioned by the Admiralty to survey the Irish

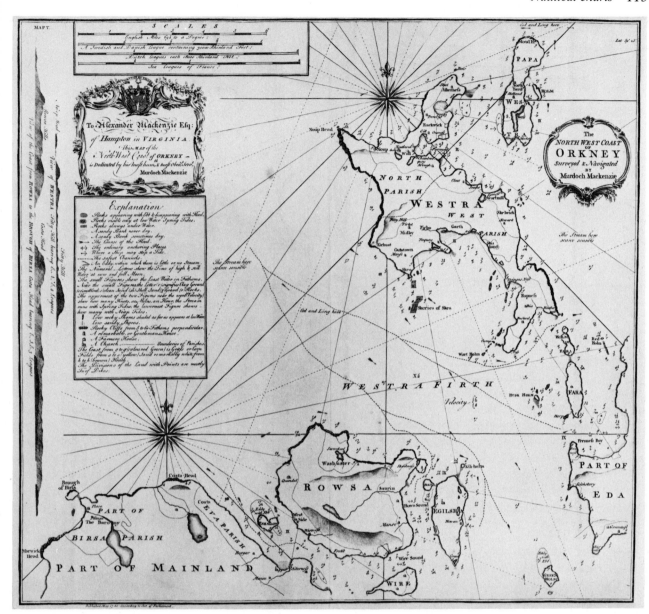

Fig 74 **Murdoch Mackenzie's survey of the north-west coast of Orkney, part of his general survey of the Orkneys, commenced in 1742, which was the earliest British marine survey to be based on scientific triangulation and a carefully measured baseline. The chart includes a lengthy explanation of conventional symbols and profiles of the coastline.**
Courtesy: **University of Leicester**

coasts and the west coast of Scotland, as a result of which he published *A maritime survey of Ireland and West Coast of Great Britain* (1776). Mackenzie described some of the commonly used methods of making a coastal survey: 'by sailing along the land, taking the bearings of the headlands by a sea compass and guessing the distance by eye or log line; . . . the common tho' less certain

method of constructing charts either without surveying, navigating, or viewing the plans themselves; but only from verbal information, copied journals, or superficial sketches of sailors'.

He indicated that more accurate surveying involved the use of a ship's log, i.e. a piece of wood dropped over the stern. The log itself was presumed to remain stationary in the water while a knotted rope attached to it was run off a reel. After a half-minute or so the rope was hauled back on board and the number of knots counted in order to assess the ship's speed. As a result the knot came to be referred to as a unit of speed equivalent to a nautical mile per hour.

Several significant advances in marine survey may be noted in Mackenzie's day: the more extensive use of

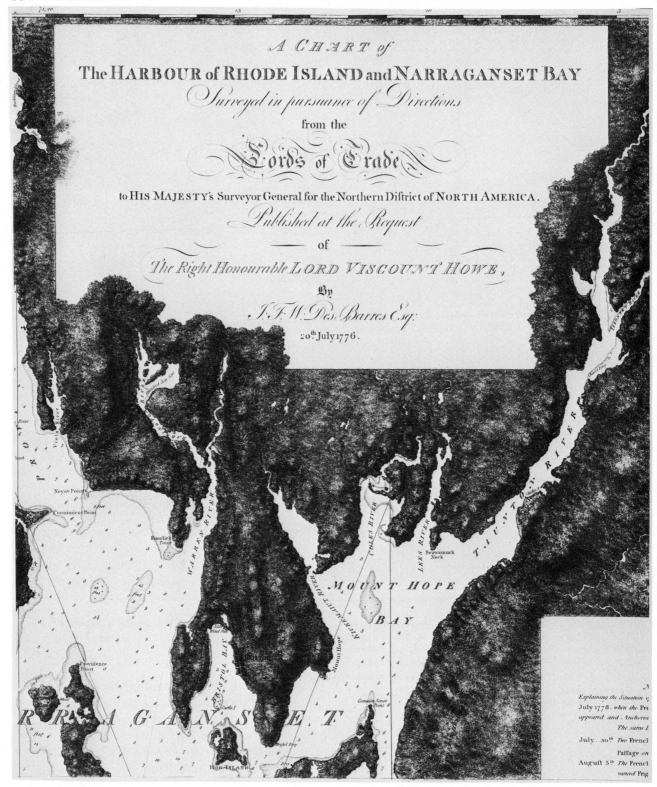

Fig 75 Chart of the harbour of Rhode Island and Narraganset Bay (1776) from *Atlantic Neptune* **by J. F. W. Des Barres. The care taken to present a vivid picture of relief is unusual in a chart designed for mariners.**
Courtesy: **University of Liverpool**

accurate triangulation; the development of better instruments for angular measurement; an increase in the number of soundings; greater accuracy in determining latitudes. Furthermore, the *Nautical Almanac* (1767) when used in conjunction with Hadley's octant, a forerunner of the sextant, made it possible to determine longitude more accurately by computing lunar distances.

'ATLANTIC NEPTUNE'

Conflict between France and England to decide which should have control over eastern North America and conflict with the Indians on all sides made it essential to have accurate charts of the eastern seaboard of North America. A programme of marine survey began under several different auspices with no overall plan; some charts were produced by private surveyors, others were supported by provincial governors, the Admiralty undertook still more. William Gerard De Brahm published the *Atlantic Pilot* (1772), containing charts of Florida, which was memorable as an early attempt to study the nature of the Gulf Stream. Samuel Holland, surveyor-general of the Northern District of North America, surveyed Prince Edward Island, the Magdalen Islands and Cape Breton Islands. Both Holland and De Brahm contributed to the outstanding success of *The Atlantic Neptune*, a superb series of some 250 immensely detailed charts, harbour plans and views, published by Joseph Frederick Wallet Des Barres, a British subject of Swiss extraction, who served as a military engineer in North America. From 1763 to 1774 Des Barres surveyed the coasts of Nova Scotia, then spent the ensuing ten years in England preparing his charts and those of his collaborators. The finished charts, designed for the Royal Navy, were often the earliest surveys of particular sections of coast and many placenames appeared for the first time to give the charts a significant historical importance. Des Barres included considerable topographical detail – fields, woods and houses as well as differences in soil structure.

The complete series covered the coasts from Cape Breton to the West Indies but a curious feature of surviving sets is that no two are identical due to the practice of assembling selected charts from the total number as sets to meet the individual needs of British naval commanders.

THE SCIENTIFIC VOYAGES OF CAPTAIN JAMES COOK

Captain James Cook, the most talented naval surveyor of his time, set new standards of exactitude in the marine surveys of unknown coasts made during his scientific voyages of discovery. The Pacific explorer and navigator entered the Royal Navy in 1755, serving in the Seven Years War and making important surveys of the St Lawrence River and parts of the coasts of Newfoundland and Labrador. His reputation as a surveyor was founded on the charts produced as a result of these surveys, but he is best known today for his Pacific voyages when he sailed under secret orders to forestall the French in the search for a Great Southern Continent, the Terra Australis Incognita. Though Cook naturally failed to discover the continent, he sailed around the entire coasts of the north and south islands of New Zealand to produce a remarkable chart in 1772 which was the result of a continuous running survey from his ship, taking compass bearings and sextant angles on coastal features and making sketches from the masthead. By the end of his voyages only a handful of Pacific Islands were undiscovered, the theory of a Great Southern Continent was disproved and New Zealand was accurately mapped.

NINETEENTH-CENTURY HYDROGRAPHY

It seems likely that George III would have given the post of Hydrographer of the Navy to Cook but for the latter's untimely death. The position went instead to Cook's sternest critic, Alexander Dalrymple, who was immediately charged with the task of compiling an inventory of information gleaned from all available plans and charts with a view to producing improved charts for use by Royal Navy commanders.

The first Admiralty chart appeared in 1801 but Dalrymple's perfectionist approach led to unacceptable delays in the production of further charts and he was unable to meet the demand. This meant a forced retirement in 1808 when he was replaced by the more practical Thomas Hurd, who quickly established a regular supply of charts to all naval stations. By 1814 survey ships had become an important part of the naval fleet and in 1829 Thomas Beaufort began a distinguished career as Hydrographer at a time of intense activity, with British survey vessels hard at work in far-flung territorial waters.

The United States established a Survey of Coasts in 1809 but little further development took place until the Board of Commissioners for the Navy in 1830 took the first step towards forming a hydrographic office when they established a Depot of Charts and Instruments. In 1832 Congress reactivated the Survey of Coasts and the United States was in the strange position of possessing two hydrographic agencies. By the end of the Civil War

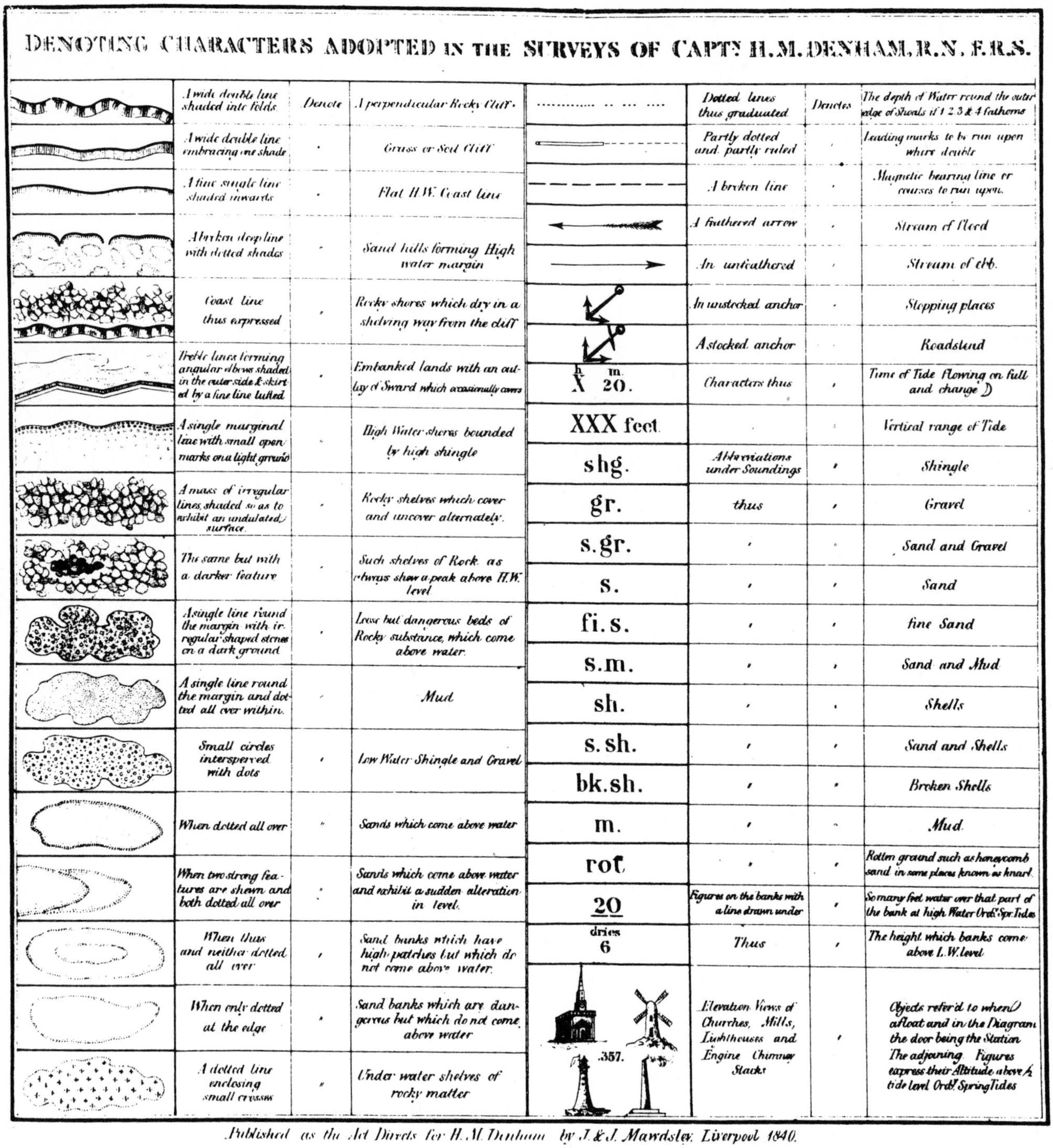

DENOTING CHARACTERS ADOPTED IN THE SURVEYS OF CAPTN H.M.DENHAM.R.N.F.R.S.

(wide double line, folds)	Denote	A perpendicular Rocky Cliff		*(dotted lines graduated)*	Denotes	The depth of Water round the outer edge of Shoals if 1 2 3 & 4 fathoms
(wide double line, one shade)	.	Grass or Soil Cliff		*(partly dotted line)*	.	Leading marks to be run upon where double
(fine single line shaded inwards)	.	Flat H.W. Coast line		*(broken line)*	.	Magnetic bearing line or courses to run upon
(broken deep line with dotted shades)	.	Sand hills forming High water margin		*(feathered arrow)*	.	Stream of flood
(coast line)	.	Rocky shores which dry in a shelving way from the cliff		*(unfeathered arrow)*	.	Stream of ebb
(treble lines, angular elbows)	.	Embanked lands with an outlay of Sward which occasionally covers		*(unstocked anchor)*	.	Stopping places
(single marginal line, open marks)	.	High Water shores bounded by high shingle		*(stocked anchor)*	.	Roadsland
(mass of irregular lines)	.	Rocky shelves which cover and uncover alternately		X m. 20.	.	Time of Tide Flowing on full and change)
(the same but darker)	.	Such shelves of Rock as always show a peak above H.W. level		XXX feet		Vertical range of Tide
(single line round margin, stones)	.	Loose but dangerous beds of Rocky substance, which come above water		shg.	Abbreviations under Soundings	Shingle
(single line round margin, dotted within)	.	Mud		gr.	thus	Gravel
(small circles with dots)	.	Low Water Shingle and Gravel		s.gr.	.	Sand and Gravel
(dotted all over)	.	Sands which come above water		s.	.	Sand
(two strong features, dotted)	.	Sands which come above water and exhibit a sudden alteration in level		fi.s.	.	fine Sand
(thus, neither dotted)	.	Sand banks which have high patches but which do not come above water		s.m.	.	Sand and Mud
(only dotted at edge)	.	Sand banks which are dangerous but which do not come above water		sh.	.	Shells
(dotted line enclosing crosses)	.	Under water shelves of rocky matter		s.sh.	.	Sand and Shells
				bk.sh.	.	Broken Shells
				m.	.	Mud
				rot.	.	Rotten ground such as honeycomb sand in some places known as knarl
				20 *(underlined)*	Figures on the banks with a line drawn under	So many feet water over that part of the bank at high Water Ord.Spr.Tides
				dries 6	Thus	The height which banks come above L.W. level
				(elevation views)	Elevation Views of Churches, Mills, Lighthouses and Engine Chimney Stacks	Objects refer'd to when afloat and in the Diagram the door being the Station. The adjoining Figures express their Altitude above h tide level Ord.Spring Tides

Published as the Act Directs for H.M. Denham by J. & J. Mawdsley, Liverpool 1840.

Fig 76 A remarkably comprehensive list of characters or symbols used by Captain H. M. Denham in his surveys of the Mersey and Dee, 1840

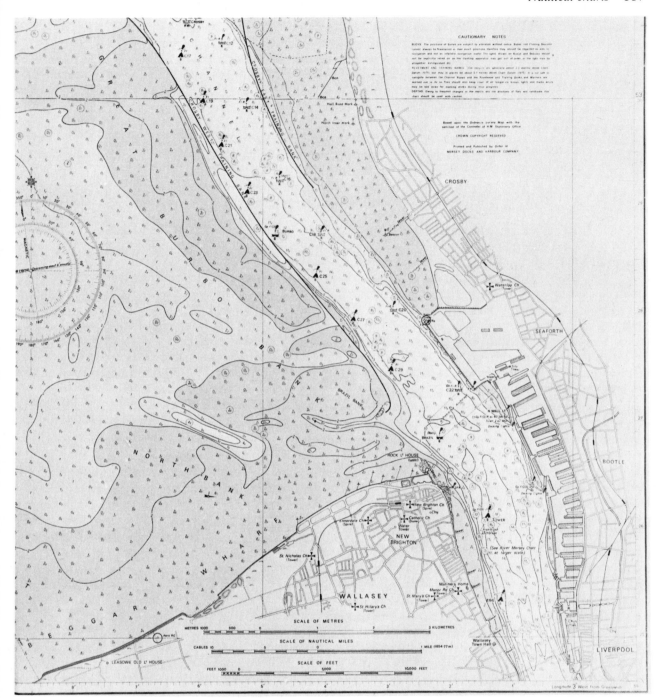

Fig 77 Detail from a modern chart of Liverpool Bay showing the channel into the Mersey prepared by the Mersey Docks and Harbour Company. Despite the vast amount of detail – soundings, banks, buoys etc. – the accent is entirely on clarity and legibility. This is of course essential in these days of huge ships and crowded waters.
Courtesy: The Mersey Docks and Harbour Company

in 1865 the Coast Survey had made great strides in the standardisation of symbols and in colour printing. Eventually the Survey of Coasts became the US Coast and Geodetic Survey and in 1866 the Naval Hydrographic Office was established in Washington. The first-named became responsible for the charting of home waters, the latter for those of other nations. Both agencies continued to exercise especial concern over standardisation of symbols and design factors such as typography, design of borders and the amount of topographic detail to be shown.

The policy of the British Hydrographic Department has always been to make their charts available to ships of any nation so that they may navigate with greater safety. Today the department publishes millions of charts, with a range of 3400, based on the work of the Navy's ocean-going and inshore survey vessels as well as on information supplied by port, harbour and river authorities, government departments and commercial establishments. Those states who are members of the International Hydrographic Bureau (established at Monaco in 1921) exchange information freely, as part of this body's avowed aim of both improving and co-ordinating hydrography at an international level.

THE MODERN CHART

The British Admiralty Hydrographic Department sets out to produce charts that meet the requirements of all users, whether they be naval or merchant seamen, fishermen, harbour masters, pilots or yachtsmen. In addition to standard charts, a number of special-purpose charts are produced to fit the needs of submarine commanders, hovercraft pilots, naval aviators and so on, while a particularly important type of contemporary publication is the Routeing Chart designed specially to help plan the routes of super tankers and other huge bulk carriers.

A modern chart must provide everything the navigator needs to know about whatever stretch of water he sails – soundings, shipping hazards, compass variations, information about currents, buoys, lighthouses and landmarks. New symbols have to be devised to meet modern developments such as oil rigs which constitute an additional hazard to shipping. Coastlines and topographical features are mapped by conventional methods of ground triangulation, combined with air survey, while hydrographic detail is amassed by survey vessels, each commanded by an officer who is as skilled in survey as in seamanship. These ships take lines of soundings by sailing up and down parallel tracks, using an echo-sounder to indicate depths, together with an electronic system of fixing position. The resulting data is interpreted and processed on board ship, then drawn up and despatched to the Hydrographic Office for fair drawing and printing. Most modern charts are prepared on Mercator's projection and appear at varying scales. Their design and format has to be geared to the limitations of a ship's chartroom; for example, chart tables are of fairly standard size and any increase in the size of charts would meet with opposition from the naval surveyors. Colours have to be carefully considered, for charts are used by night as well as by day and the lighting on a ship's bridge is subdued. Amber, red and yellow lamps are used and any colours to be incorporated in a chart have to be subjected to rigorous tests under these lighting conditions to ensure that colours are clearly differentiated. Charts have generally been printed in three colours only: black for detail, blue tint for shallow waters and magenta for lights. On more recent charts, however, a buff tint is used on land areas and adds a great deal to the overall clarity. Symbols have been standardised by the International Hydrographic Bureau and cover a wide range of hydrographic and topographic phenomena. One significant recent development is the introduction of metrication, with depths given in metres rather than fathoms and feet, but the day is still somewhat distant when metrication will be applied to the entire range of Admiralty charts.

Today's chart is a worthy culmination to almost 700 years of endeavour and research by the cosmographers, cartographers, mathematicians and hydrographers of many nations. To the skilled navigator with his sophisticated instruments, his navigational tables and above all his accurate chart, the sea is as well marked as any motorway and even the spare-time small boat sailor can navigate confidently along an unfamiliar coastline in a manner unknown to the mariners of days gone by.

8 ROUTE MAPS

One of the fundamental purposes of maps throughout cartographic history has been the indication of routes from one place to another. Networks of communication have not only figured prominently on topographic maps but an ever-increasing output of specialist maps has been devoted to route maps of various kinds. Each new development in modes of travel – by sea, road, canal, rail, air – has involved the production of many new maps, with the mapmaker concerned at all stages of the project from its conception to its completion. At the planning stage of a proposed motorway, for example, the mapmaker has to update and revise existing maps as well as to make fresh surveys, and later when the route is fully operational he has to supply maps to meet the needs of those who will use it.

ROAD MAPS

The previous chapter was concerned with the long, slow development of nautical charts. This one will be concerned with land and air routes and it is appropriate that a beginning should be made with one of the oldest and most widely-used types of specialist map – the road map. At a time when Rome was occupied with the administrative and military requirements of a great empire her cartography was of an essentially practical nature. The Romans had little time to concern themselves with the scientific theories and speculations of Greek scholars, and concentrated on surveys which would assist their administrators and facilitate the movement of military and commercial traffic. It is known from literary references that a plentiful supply of maps was available in ancient Rome but very few have survived. Road surveys of the Empire were initiated and the results incorporated into practical route maps. One example was the map made by Marcus Vipsanius Agrippa and though no copies survive there is speculation that the most important extant Roman map, known as the *Tabula Peutingeriana* or Peutinger Table, may have been based on Agrippa's work.

The Peutinger Table

Visual evidence of the nature of a Roman road map is provided by the Peutinger Table, a work which survives remarkably after a chequered history. Bagrow states that a copy was made *c.* AD 250 of an original route map made in the first century AD (*History of cartography* (London, 1964), pp. 37–8); one hundred years later improvements were effected to the delineation of coastal areas and additional islands located; in *c.* AD 500 further improvements were made, and minor corrections followed in the eighth and ninth centuries. The much revised original map is now lost but by a stroke of good fortune a copy was made during the eleventh or twelfth centuries at an unspecified library in southern Germany. This copy was located in 1507 by a Viennese, Konrad Celtis, who later bequeathed it to the Augsburg collector, Konrad Peutinger, after whom it is now named. The involved story of the map does not end even here, however, for after Peutinger's death in 1547 the map was lost until 1591 when the Burgomaster of Augsburg, Marcus Welser, found part of it and had it engraved and published. Seven years later the entire map was recovered and taken to the Antwerp cartographer, Ortelius, who had it engraved for publication.

What of the map itself? It is diagrammatic in construction, owing no allegiance to any map projection and not drawn to any constant scale. It is produced on a roll twenty-one feet long by only one foot deep, a format which resulted in acute distortions of shapes and spatial relationships. Nevertheless it had a practical purpose in that it formed a light, compact roll which was easily transportable for field use. The map detail was entirely functional with a concentration on the communications network of the Empire and the highway distances between cities and military garrisons. The roads were delineated as thin red lines, while important places, trading centres, mineral springs, ports and harbours were illustrated pictorially in perspective view. The hub of the Empire, the city of Rome, was symbolised by a pictorial

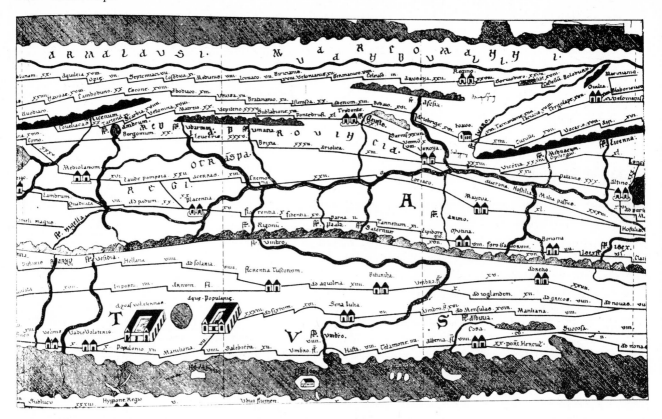

Fig 78 **Detail from the *Tabula Peutingeriana* or *Peutinger Table*, a road map of the Roman Empire which came into the possession of the Nuremberg scholar, Konrad Peutinger, after whom it is named. The map is grossly distorted owing to its peculiar dimensions, 21ft by 1ft. The detail shows northern Italy, the main river shown being the Po. Towns such as Mantua, Cremona, Placentia and Florence can be picked out. The illustration appears in Konrad Miller's *Itineraria Romana* (Stuttgart, 1916).**
Courtesy: **University of Liverpool**

personification contained within a circle, from which twelve named highways radiate.

Medieval route maps

In medieval times the requirements of military strategy and imperial administration were subordinate to the needs of pilgrims wending their way to the religious centres of Europe and the Holy Land. Accounts of such journeys appear in early itineraries such as Bernhard von Breydenbach's *Peregrinatio in Terram Sanctam* (Mainz, 1486), a work which included a map of Palestine and views of cities.

An earlier itinerary which had greater significance in cartographic development is contained in *Chronica Majora* (1259) by the monastic historian of St Albans, Matthew Paris, who provided a pictorial route map in

vertical strips to illustrate the pilgrim route from London to the Holy Land, via Boulogne, Paris, Rome and Otranto. In using the strip technique Matthew Paris pioneered a method used with great success by the seventeenth-century English mapmaker, John Ogilby, and which remains popular today. Matthew Paris catered for the needs of pilgrims by locating important halting places on the route, depicting them by a stylised drawing of buildings.

Matthew Paris is more widely known for four versions of a map of Britain in which attention is focussed on the pilgrim route from Newcastle to Dover, the port of embarkation for Europe. In the best of these maps Matthew Paris distorted the shape of Britain in order to display the pilgrim route vertically down the centre of the map. Although he lacked sufficient space to delineate the road itself, its course can be easily followed from the many settlements, ninety per cent of them monastic sites, which are located along it. Despite numerous erroneous spatial relationships the map gives a recognisable picture of Britain, in marked contrast to the much inferior configuration of the slightly later Hereford Map. Paris's map was oriented to the north – in contradiction to general medieval practice – and he showed the four cardinal points. No scale of distance was provided, though Matthew Paris must have been conversant with the distances between places for, in his compilation, he must have frequently resorted to reports from travellers, merchants and monks from other monastic houses.

The Gough or Bodleian Map

The remarkable map known either as the *Gough Map* (after its discoverer, Richard Gough) or the *Bodleian Map* (from its resting place in the Bodleian Library, Oxford) was prepared *c.*1360 by an unknown maker. It represented a major advance in British mapmaking, particularly for its illustration of the contemporary road network. Despite subjecting Scotland to a north-south elongation and an east-west contraction, the mapmaker achieved an instantly recognisable and surprisingly accurate outline of the British coasts and it is surmised that the configuration was based on a framework supplied by a plot of the main roads. Richard Gough defined the map's greatest merit to be 'that it may justly boast itself the first among us wherein the roads and distances are laid down'. It is believed that the map was intended for practical use by travellers and may have been used by government couriers and other servants of the Crown. No attempt was made to plot the actual course of the roads, which appear as straight, red lines with distances indicated in Roman numerals. It is unlikely that these distances were the result of deliberate measurement from one place to another but were based on the most reliable reports available to the author. Sir Charles Close calculated from eight distances on the map that they provide evidence of a mile of 2335 yards; Gough himself estimated 2358 yards, while the Old English Mile in use at the time has been calculated, from the distances between certain milestones in Yorkshire, to be 2424 yards. Irrespective of the methods used in assessing distances, the remarkable way in which they tally with many of the mileages of Ogilby's *Britannia*, made three centuries later in 1675, provides some testimony to their accuracy.

Early European Itineraries; Mexican Road Maps

Roadbooks and itineraries survive from an early date in France, where the network of post-roads, together with a regular supply of post-horses and guides for the aid of travellers, dates from an *Ordonnance* of Louis XI in 1464. Several itineraries printed in the early sixteenth century described places encountered and indicated the distances between halts on pilgrimage routes to Rome, Jerusalem and Santiago de Compostela. It is assumed that such works made some impression in Britain and may have influenced the preparation of early English itineraries.

The first road map in printed form was a woodcut made in Nuremberg *c.*1500 by Erhard Etzlaub. This *Romweg* map served as a guide for pilgrims journeying to Rome in Holy Year and displayed the best routes through Germany. The map appeared as a broadsheet and on it Etzlaub used the convention of small circles to symbolise towns and drew dotted lines to show the most suitable linking routes between them.

In Antwerp, Johannes Metellus Sequarus prepared *Itinerarium orbis christiani* (1579–80), a work which is thought to be the first printed road atlas and which was a pioneer in using the convention of a double parallel line to represent roads.

Mapmakers generally adopted some kind of linear convention to indicate roads and while it is absolutely logical to represent a linear feature on the ground by corresponding lines on a map, Mexican mapmakers in *Codex Tepetlaoztoc* (mid sixteenth century) combined this type of European symbolism with rows of footprints to represent tracks, thus providing a very evocative symbol, though it lacked the precision of linear representation.

Descriptions of Roads in Sixteenth-Century Britain

The *Itinerary* of John Leland, the *Chronicles* of Raphael Holinshed and William Smith's *Particular Description of England . . . 1588* provided written information about the countryside traversed but, with few exceptions, the cartographic display of roads was neglected. Norden inserted roads on his county maps and introduced the triangular distance table, a device which efficiently supplies a remarkable amount of information in a compact space.

John Ogilby's 'Britannia'

A major advance came with Ogilby's *Britannia* (1675) in which the principal roads of Britain were presented in strip form on 100 plates, six or seven strips to each plate. *Britannia* set a pattern for route delineation which, like the distance table, remains as effective as ever today. It was important on several counts; it was based on the first systematic survey of British roads; it introduced the statute mile of 1760 yards and eliminated the practice of using local 'customary' miles; it established the one-inch-to-one-mile scale as a standard; and it exerted such an influence on British mapmaking that, as far as roadbooks were concerned, only derivatives appeared until John Cary made a new road survey at the end of the next century. The strip method is a highly efficient method of communicating information to a road traveller for it concentrates his attention on the route itself, with no extraneous detail away from the road to divert him. He is

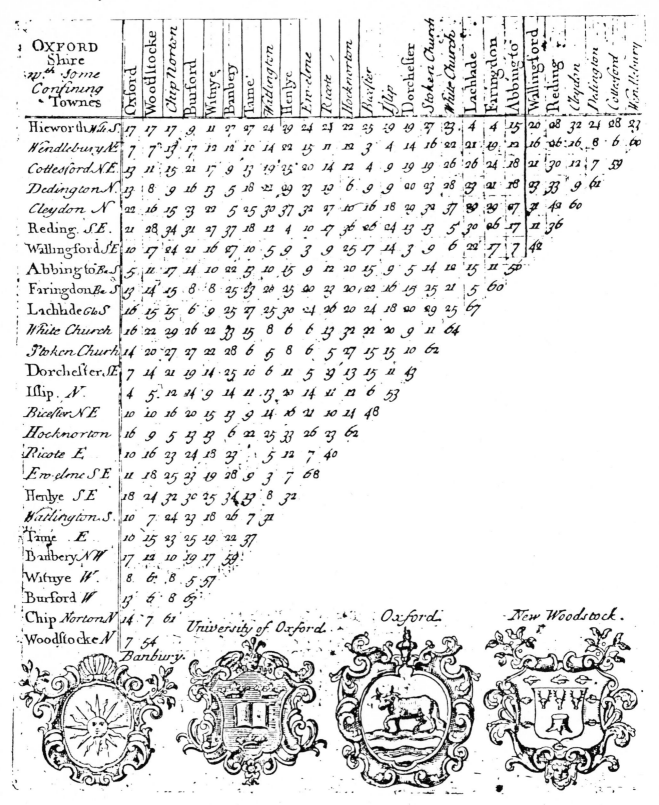

Fig 79 **Triangular distance table from Robert Morden and Thomas Cox's** *Magna Britannia* **(1738). The distance table, a useful device for referring to distances between towns, was first used in Britain by John Norden in** *England An Intended Guide, For English Travailers* **(1625)**

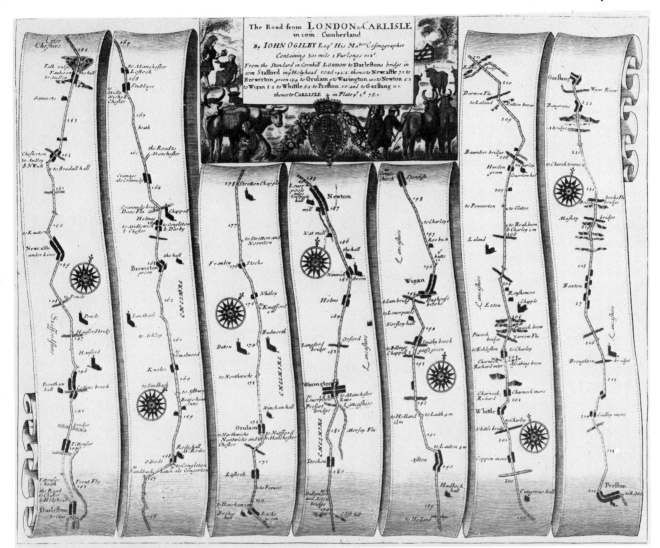

Fig 80 Strip map from John Ogilby's *Britannia* (1675) showing the Darlaston to Preston section of the London-Carlisle road. Compass roses are inserted as an indication of direction and the dots along the roads indicate the divisions of the miles into furlongs.
Courtesy: University of Liverpool

given just enough information to make his journey interesting as well as to facilitate his progress. Ogilby explained the method of using his maps in his Preface:

> The Initial City or Town being always at the bottom of the outmost Scroll on the Left Hand; whence your Road ascends to the Top of the said Scroll; then from the Bottom of the next Scroll ascends again, thus constantly ascending till it terminates at the Top of the outmost Scroll on the Right Hand . . . the Road itself is express'd by double Black Lines if included by Hedges, or Prick'd Lines if open . . . The Scale . . . is according to one Inch to a Mile . . . the said Miles being exprest by double Points, and numbered by the Fig-

ures 1, 2, 3 &c. Each subdivided into 8 Furlongs, represented by Single Points. Ascents are noted as the Hills in Ordinary Maps. Descents e contra, with their Bases upwards.

From *Britannia*'s time forward it became a commonplace to include roads on English county maps. In 1676 Robert Morden issued two sets of playing cards bearing county maps with roads based on Ogilby. These were the first county map series to include roads, though their miniature size and unconventional format means that this distinction is usually given to Morden's set of county maps which illustrated Gibson's translation of Camden's *Britannia* (1695).

Roads appeared on other types of British map as well as county maps and strip maps. They were a prominent feature of the large wall maps and sheet maps which were common during the seventeenth century. John Adams's twelve-sheet map of England and Wales (1677), for example, is distinctive for its framework of straight lines linking towns, the distance between neighbouring places being noted. Smaller maps too, such as Philip Lea's *A Travelling Mapp of England Containing the Principal Roads* (*c*.1695), delineate the major roads and give the mileages along each section.

Ogilby's *Britannia* was much less practical, for its bulk meant that it was far more suited to the prior planning of a journey than for use *en route*. Despite this obvious disadvantage, forty-four years elapsed between *Britannia*'s publication and that of a more compact volume. In 1719 and 1720 three smaller derivatives of Ogilby appeared. The first was Thomas Gardner's *A Pocket Guide to the English Traveller* (1719). In his preface Gardner put the case against Ogilby succinctly. 'As the original Plates are in large Sheets, the general Use of them has been hitherto lost, and the Book rather an Entertainment for a Traveller within Doors, than a Guide to him upon the Road.' In 1719 also, the established mapmaker John Senex issued *An Actual Survey of all the Principal Roads of England and Wales* and in 1720 the distinguished engraver, Emanuel Bowen, prepared strip maps and John Owen compiled the text for *Britannia Depicta or Ogilby Improv'd*, a work which included small county maps in addition to the road strips.

The rapidly increasing demand for road maps led to the first road atlases of Scotland and Ireland: the *Survey and Maps of the Roads of North Britain, or Scotland* (1776) and *Taylor and Skinner's Maps of the Roads of Ireland, surveyed 1777*, both by George Taylor and Andrew Skinner. The authors state that 'the Military Roads are kept in the best Repair; and so much has been done of late years to the other Roads . . . that Travelling is made thereby incredibly easy, expeditious, and commodious.' In 1775 R. Sayer and J. Bennett published *Jeffery's Itinerary; a Traveller's Companion*, using the strip map technique and giving its raison d'être as due to 'the many new Turnpike Roads . . . few of which are taken notice of in any of the books of the English Roads'. It was, however, John Cary who made the greatest eighteenth-century contribution to British road mapping when he surveyed the whole English road system as the result of a commission from the Surveyor and Superintendent of the Mail Coaches, General Post Office. Cary's publications included *Cary's Traveller's Companion*, which was not a roadbook in the true sense but 'a Delineation of the Turnpike Roads of England and Wales . . . Laid down . . . on a new set of County Maps'. *Cary's New Itinerary* (1798), a very successful roadbook, achieved eleven editions despite strong competition from the rival roadbook of Daniel Paterson, a work in which tables of distances along a route were supplemented by copious descriptive notes and general

maps. The finest work of this period, however, was *Cary's New Map of England Wales* (1794) in eighty-one sheets at a scale of five miles to one inch. This work was

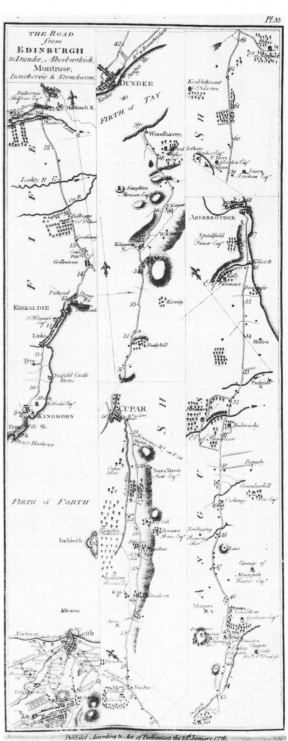

Fig 81 **Strip map by the Aberdeen surveyors, George Taylor and Andrew Skinner in** *Taylor and Skinner's Survey and Maps of the Roads of North Britain or Scotland* **(1776).**
Courtesy: **University of Liverpool**

based on Cary's Post Office road survey and its accuracy and fine engraving made it the most popular map with travellers of its day.

Three main techniques of recording roads have been noted; first, the strip map method; secondly, distance tables accompanied by descriptive notes; thirdly, county and regional maps in which roads were the prominent feature. *Laurie and Whittle's New Traveller's Companion* (1806) deviated slightly from these standard patterns in that it comprised an atlas of regional maps so arranged that the direct route from one place to another could be followed on successive maps.

The automobile map

Road maps today are geared almost exclusively to the motorist but the relatively recent development of the motor car has meant that automobile maps occupy only a short period in the story of cartographic progress, being largely a twentieth-century product. Their significance and usefulness, however, is out of all proportion to their restricted history and without doubt the automobile map is one of the most important American contributions to cartography, for it has been an indispensible facet of everyday life in the USA since the origination of the American automobile industry in 1895. Ristow tells how in that year the American automobile map was born when the Chicago *Times-Herald* prepared a map showing the course of a fifty-four mile race in which Frank Duryea set gasoline cars against other self-propelled vehicles in a successful attempt to demonstrate their superiority (*Surveying and Mapping*, Vol. XXIV No. 4 (Dec. 1964), p. 617). Before this time American cyclists had fostered an enthusiastic campaign for better maps as well as for improved roads. Their efforts met with considerable success, for several excellent roadbooks and maps designed exclusively for cyclists were published. These did not, however, meet the needs of the new breed of motorists whose vehicles required wider roads than the narrow tracks which sufficed for the cyclists.

In 1901 the first *Official Automobile Blue Book* featured four road maps as a supplement and further works of a similar nature appeared in the early years of the twentieth century. Despite a brief interest in photo-motoring guides, in which photographic views of the routes were supplemented by written text and small maps, no significant developments took place until the Gulf Refining Company set up its first filling station in Pittsburgh in 1914. This was the first of many, and one of the services offered by these stations was the distribution of free road maps, the brain-child of a Pittsburgh advertising man, William B. Akin. The idea caught on rapidly,

production gained momentum and major map publishers such as Rand McNally became involved in the preparation of oil company maps. For a brief period in the 1920s the Michelin Tyre Company set up a cartographic business in New Jersey and though this was not economically successful, the Michelin maps introduced new standards and made a significant contribution to the subsequent improvement in American road maps. Although the automobile map, unlike most other map forms, is essentially American in origin, the motor car also caused a revolution in mapmaking in Europe where Michelin, in 1910, began their popular series of guide books, followed by the sale of maps in 1913.

During the inter-war period three major cartographic producers for the oil companies – General Drafting, H. M. Gousha and Rand McNally – issued some 150,000,000 maps per year for distribution through petrol stations. Today's production is immense, and since World War II oil companies in Britain and Europe have issued millions of maps to motorists – not usually gratis as in the US but at a very modest price. European and American companies have kept up with the burgeoning demand from vacationists taking overseas motoring holidays by producing not only automobile maps of their native territories but also of overseas holiday areas. The quality of British road maps and atlases is high due to the participation of such establishments as Philip and Bartholomew, and excellent maps are published in Europe by Kummerley and Frey, Hallwag, Michelin and others. Competition is naturally intense to produce clearer, more aesthetically pleasing, more useful and more up-to-date maps. To this end the commercial makers of national road maps regularly consult with their relevant transport ministries before compilation or revision takes place. Surveys from the air and on the roads themselves are highly important, and the Michelin company has introduced a new technique of information gathering in which information is secured about roads shown on its maps by recording a verbal commentary while driving along the route and later transcribing it at headquarters.

The primary function of any automobile map is limited to providing the motorist with relevant information about his journey and in so doing the map repeats a certain amount of information which automatically occurs on a standard topographical series. Indeed the O.S. quarter-inch-to-one-mile map was long regarded as the motoring map *par excellence*. The purpose of the specifically-designed automobile map, however, is to concentrate the motorist's attention on his route and it therefore omits a good deal of topographical data, particularly about relief, and does not by any means replace the topographical map as a general-purpose source of information.

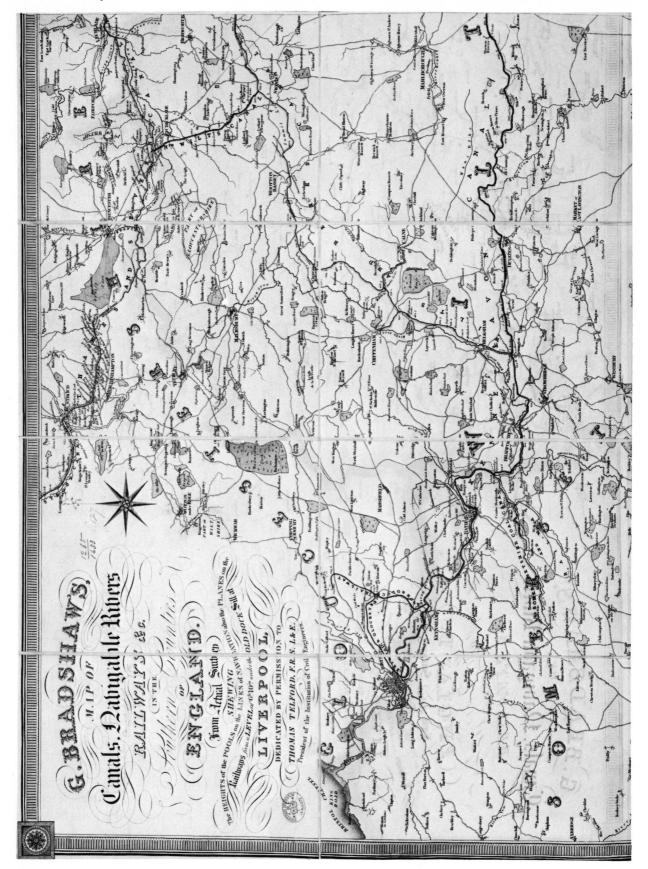

WATERWAY MAPS

As a primary means of communication, waterways received considerable attention from early mapmakers and were often the most eye-catching features of an early map. In China waterways have always been of major importance in the social and economic system and as long ago as the first century BC a volume entitled the *Shui Ching* or Waterways Classic was written. Much later Shan O spent over thirty years exploring the lakes, rivers and canals in the region of Suchow, Changchow and Huchow and consequently prepared a treatise entitled *Wu Chung Shui Li Shu* (The Water-Conservancy of the Wu District) in AD 1059. The *Hsing Shui Chin* (Golden Mirror of the Flowing Waters) issued in 1725 included many panoramic maps of rivers and lakes executed in a particularly delicate pictorial style with buildings, hills and trees seen in perspective and rivers filled in with a flowing pattern to suggest direction of flow. These panoramas, quite unlike anything produced in Europe because of their affinity with Chinese landscape painting, provide a vivid picture of river and lake topography.

In Europe medieval mapmakers such as Matthew Paris gave great prominence to rivers, though he was not averse to distorting their course in order to accommodate his pilgrim routes more satisfactorily. Saxton also featured rivers prominently on his county maps and was careful to indicate important crossing points which were vital to travel in the late sixteenth century. The first canal to be cut in modern Europe was the Languedoc Canal, cut between 1667 and 1681 from the Mediterranean to Toulouse, where it joined the Garonne which could be followed to Bordeaux, thus providing a short through route by waterway to the Atlantic. This canal was illustrated on a large three-sheet map by J. B. Nolin in 1697, the map itself being surrounded by plans of the forty-five aqueducts encountered in a traverse of the canal. In Britain river navigations such as the Severn carried heavy traffic in the seventeenth century but it was not until after 1761 when the Bridgewater Canal was opened that the entire system of British canals began to evolve. Thenceforward the rapid development of waterborne transport meant that there was a satisfactory alternative to the poorly-built or neglected road system of the time. The development of new modes of transport

invariably necessitates new surveys and up-to-date maps. Those prepared in connection with river navigations varied from small-scale works showing little more than the general line of the navigation to large-scale plans prepared by the great civil engineers – James Brindley, Thomas Telford, John Rennie, John Smeaton and so on. The same was true of canal maps which became a vital part of the mapmaker's stock-in-trade; John Cary, for example, was responsible for numerous canal, dock and drainage system maps and plans, varying in scope from a single sheet plan of the navigable canal system (1779) to detailed plans of proposed docks in London (1796) and a plan of lands at Paddington, proposed to be leased and exchanged by the Bishop of London and his lessees and the Grand Junction Canal Company (1812). He also produced an atlas of canal plans entitled *Inland Navigation; or Select Plans of the Several Navigable Canals throughout Britain*. Dedicated to the Duke of Bridgewater, this work included sixteen folding plates at half-inch to one mile, each plate showing county boundaries, topographical detail and the existing and proposed lines of canals. This new form of transport called for the introduction of additions to the cartographic language to represent locks, bridges, tunnels, aqueducts and so on. The new conventions affected county maps as well as special purpose maps of canals so that on Burdett's map of Cheshire (1777) the now-familiar arrow-head symbol for locks is seen on the Weaver Navigation and a tunnel at Preston Brook is represented by parallel broken lines.

Large-scale, accurately surveyed plans were deposited for the purpose of obtaining the Parliamentary Act necessary for construction of any canal, and were made before construction, and whenever any alteration or addition was made to a section of canal. These plans were entirely concerned with land in the immediate vicinity of the line of a canal and showed the terrain in some topographic detail. In Scotland, for example, the Forth and Clyde Canal was begun in 1768 but after a very short stretch was completed funds ran out and a ten-year delay ensued before work recommenced with Robert Whitworth as engineer. On his authority a fine plan was made from a survey by John Laurie. The plan includes part of the shallow channel of the Clyde, the depth being indicated by soundings in feet. Short walls had been built out into the river in an attempt to deepen the river by confining it to a narrow, faster-flowing channel. These walls appear on the plan as short parallel lines. Numerous locks are located on the canal itself by the double arrow-head symbol and some indication of the topography is given by hachuring.

Large numbers of general maps were also made to show the British canal system as a whole, occasionally

Fig 82 Detail from George Bradshaw's Map of Canals, Navigable Rivers, Railways etc. in the Southern Counties of England (c.1830), one of a series of more reliable depictions of the mid-19th century communications network in Britain.
Courtesy: British Library Board   Maps 1.b.22

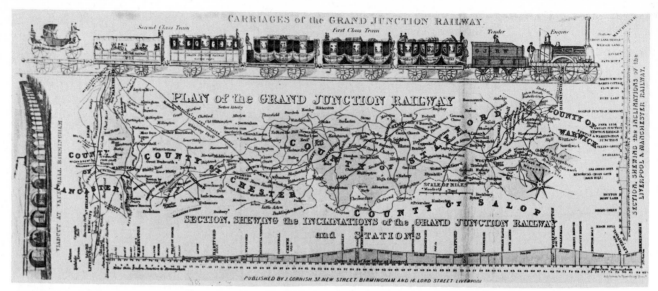

Fig 83 **Plan and section of the Grand Junction Railway from** *Cornish's Guide and Companion to the Grand Junction and the Liverpool and Manchester Railways* **(3rd edition, London, 1838).**
Courtesy: **J. E. Vaughan**

accompanying reference works such as Joseph Priestley's *Historical Account of the Navigable Rivers and Canals of Great Britain* (1831). Although these maps are useful historical evidence they should be approached with caution as they occasionally delineated canals and railways which had been authorised but were not necessarily built. Thomas Moule's *Map of the Inland Navigations of England and Wales* (1836) is a case in point, for Moule relied on the evidence of deposited plans rather than on a first-hand check of actual existence. Infinitely more reliable were the maps of George Bradshaw, a name popularly associated with railway timetables. Bradshaw's maps showed existing canals accurately and located wharves, tunnels and locks. A particularly useful item for the boatman was the indication of the length of each tunnel.

In North America great numbers of maps of canals and rivers were published during the late eighteenth century and throughout the nineteenth century. Some of these had a military function; the *Plan of the Narrows* in the East River, N.Y., made by an anonymous mapmaker in 1776 at a scale of 500 feet to an inch, showed not only the navigable channel, shoals and depths of water but also the various batteries proposed on each bank to prevent an enemy sailing up to New York; Des Barres in his *Atlantic Neptune* included a map of part of the Hudson River showing the position of Fort Montgomery and Fort Clinton, along with various cables and chains,

which were so placed as to obstruct the passage of His Majesty's forces up river. Several maps were made of the Sabine River in the 1840s to demonstrate its function as the boundary between the United States and Texas laid down by survey in 1840. Other maps prepared by direction of canal commissioners from engineers' surveys such as *A new map and profile of the proposed canal from Lake Erie to Hudson River* (1821) contrasted with purely tourist maps such as *The Hudson by daylight map* (1878) which named streams and islands, gave the heights of mountains and the names of prominent residences.

Today's maps of waterways largely serve an entirely different function from those of the nineteenth century, for they are generally designed to assist the progress, not of commercial traffic, but of the holidaymaker in his motor cruiser or converted narrow boat. Often produced in strip map form, such maps provide detailed information about everything encountered on a passage through a particular canal – locks, bridges, filling stations, canal-side public houses, tunnels and items of scenic interest.

RAILWAY MAPS

Despite the threatening dominance of official bodies such as the Ordnance Survey, the nineteenth century was a remarkably busy time for the private surveyor and mapmaker. Not only was he increasingly occupied with the production of general purpose maps and with estate and tithe surveys but he had also to meet the requirements of three competing transportation systems – roads, canals and the new railways.

The provision of railway maps did not necessitate any drastically new techniques of graphical communication, though a number of new symbols had to be devised. Those innovations which did take place related to greater precision in surveying and levelling or to better methods of map reproduction. The pattern of railway mapping was similar to that followed for canals; any proposals to build a new railway or to make alterations or additions to an existing one required accurate surveys, followed by the depositing of detailed plans, often supplemented by accurately-levelled longitudinal sections. Such deposited plans had to be sufficiently viable to withstand the challenge of vociferous opposition, for not everyone welcomed the coming of the railways. British railways were first delineated on a printed map in Vandermaelen's *Atlas of Europe* (Brussels, 1829–33) but the first truly specialist British railway map appeared in 1830 when George Bradshaw published a map devoted to the navigable rivers, canals and railways of the Midlands. He followed this with a map of the southern counties and in 1847 published a railway map of the whole of Britain. At the same time private mapmakers were adding railways to county maps and the Ordnance Survey was doing likewise to the sheets of the one-inch Old Series, particularly to the electrotype printings issued after 1847.

Among the most reliable sources of information about nineteenth-century railways in Britain were the maps of Zachary Macaulay and John Airey, both employed by the Railway Clearing House, a body inaugurated to ease the booking and general movement of rail passengers and freight and to act as a central body which would apportion payments among the railway companies concerned. Macaulay's *Station Map of the Railways of Great Britain* enjoyed such popularity that at least twenty-one editions appeared between 1851 and 1893. In 1869 John Airey began publication of a series of sectional railway maps, printed in colour by the new lithographic process. Airey used colours to differentiate the lines of the various companies and precisely indicated the distances between stations, sidings and points.

Transportation mapping in Europe and in the USA paralleled that of Britain, beginning with the production of road maps initiated by the construction of toll roads, proceeding to maps concerned with canal construction and culminating with the mapping of the steam railroads. By 1860 in the United States, railroads were playing a very important part in the life of the country; lines were constructed to link commercial centres; the Civil War provided an incentive because of the importance of rail transportation to the armies; after the war the tourist potential of the railroads was realised and a route was postulated in connection with the opening of the Yellowstone National Park. As far as railroad maps were concerned, improvements in papermaking and

printing, particularly the introduction of lithography, contributed significantly to the quality and number of maps issued. The New York publisher, Joseph Hutchins Colton, published some of the finest American nineteenth-century commercially-produced railroad maps and it is estimated that he was responsible for around thirty per cent of the commercial output. Among the Library of Congress Map Division's collection of over 5000 railroad maps are government surveys, promotional surveys, maps showing the extent of railroad land grants, progress report surveys and commercially-produced route guides. Land grant maps were often used by land speculators to publicise railroad lands for public sale. As early as 1868 western railroads established profitable land departments, with European offices, to sell land and promote foreign settlement in the western states. Modelski writes that maps were produced in foreign languages and aimed at immigrants already settled on the east coast and at prospective emigrants from Europe (*Railroad maps of the United States* (Washington D.C., 1975), p. 9). Some of these maps were deliberately distorted to present an enticing picture of one state, area or line to the advertiser's advantage. The practice of manipulating scale, area and routes was common in advertising maps of the 1870s and 1880s.

The production of railway maps continued unabated into the early years of the twentieth century when the networks were completed. Production declined from this peak, particularly in relation to promotional maps which were no longer required. Today railways play no major part in cartographic output though small railroad atlases occasionally appear. Maps are used, frequently, in railway advertising and mention should be made of the decorative maps of the London Underground System prepared by F. Macdonald Gill (1884–1947) which were among the first modern decorative maps. The stylised map of the Underground system, which must have been consulted by more people than any other single map, was unique when it first appeared, but similar maps have been produced for the Paris Metro and the New York subway, and British Rail are using a similar technique in their publicity maps for different regions and conurbations.

AIR MAPS AND AERONAUTICAL CHARTS

From the time when man took to the air, initially in hot air balloons, the need arose for a new form of map specially designed to suit the requirements of the aviator looking down on the land instead of merely travelling across it. It is not surprising, therefore, that the pioneer of air mapping, Hermann Moedebeck, was himself a balloonist. In 1888 Moedebeck made his first plea for

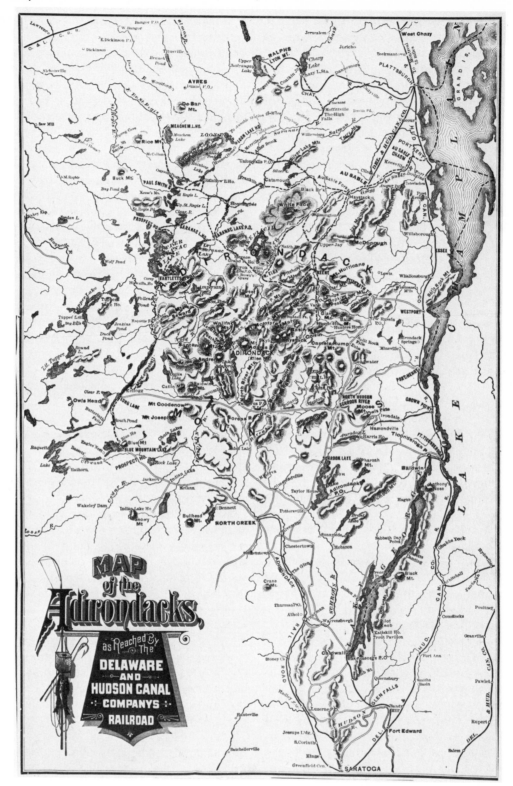

Fig 84 **A map from** *The Englishman's Guide Book to the United States and Canada* **(1884 ed. London and New York) which demonstrates how the advent of the railroad has made the Adirondacks more accessible to the traveller.**
Courtesy: **J. E. Vaughan**

good charts which would be marked with suitable landing points and help the aviator identify different types of terrain from the air. In 1906 a second attempt to arouse interest in air maps bore little fruit, but in 1907 Moedebeck was appointed Chairman of the newly-established International Commission for Aeronautical Charts. In this new role he proposed that as an initial step each nation should overprint aeronautical symbols on to existing maps and suggested suitable symbols for the purpose. In 1909 the German Aeronautical Society sponsored the publication of a map of the Cologne area at 1:300,000. This was the first true air map and was the first of a projected series.

The change from ballooning to powered flight gave a vigorous impetus to air mapping, and a meeting of the International Commission in 1911 earnestly considered how to provide better maps for the new breed of aviators. France, Germany and Austria began to produce maps quite rapidly but the first British map produced specifically for airmen did not appear until 1912. This was a map prepared for use in an air race around the London area sponsored by the *Daily Mail*. The earliest US air map was published a year earlier. It was a map of the western part of Long Island illustrating the location of aerodromes and the best air routes. In these early days it should of course be remembered that flights were at relatively slow speeds and of short duration.

The outbreak of World War I caused an accelerated demand for air maps. Only France of the combatant nations had completed preparation of even a reasonable number of sheets of a basic air map series. Early British fliers managed as best they could with standard Ordnance Survey sheets and, despite the rapid developments in aviation technology, air maps were generally derived from existing topographic material. After the war, however, specialised maps designed for commercial air traffic began to be issued, and in 1921 the First International Congress on Aerial Navigation noted the necessity for standard maps at several scales which would meet the needs of aviation planning, navigation on short and long range flights, and landing procedures.

The strip map technique, long applied to roads, was used for some air routes during the 1920s. It had peculiar advantages for aviation due to the difficulties pilots experienced in using large sheet maps in the restricted space of their cockpits. With the introduction of larger planes the strip maps became less popular and were abolished in the USA in 1935 in favour of the US Sectional Aeronautical Chart. The inter-war period saw significant advances in air map development and around forty air map series relating to various countries were available.

The onset of World War II meant the temporary closure of most commercial international flights at the very time when scheduled trans-Atlantic flights were becoming established. Military exigency, on the other hand, meant that all resources were harnessed to the production of air maps for military needs on a world-wide scale. Millions of aeronautical charts were produced between 1939 and 1945 and it was in fact during this period that the term 'aeronautical chart' became accepted, to accord with the dictum that the word 'chart' is normally associated with aids to one form of navigation or another.

Post-war developments in aviation were dramatic with faster and faster aircraft coming from the production lines and air space becoming increasingly congested. Events necessitated a re-appraisal of the whole air chart situation and in 1947 the International Civil Aviation Organisation (ICAO) was placed on a permanent footing, with its Aeronautical Charts Division charged with the responsibility for standardisation of air charts and the supervision of the preparation of a world series, the *World Aeronautical Chart* (WAC), at the one-millionth scale.

At the present time different types of aircraft and navigation necessitate charts at varying scales: 1:250,000 for helicopter pilots; 1:500,000 and 1:1,000,000 for visual navigation; 1:1,500,000 for radio navigation; 1:2,000,000 for jet aircraft; 1:5,000,000 for global flights. Those charts which are prepared specifically for visual navigation naturally feature topographical as well as aeronautical data, while those designed for instrument navigation emphasise navigational aids. Visual navigation of course means that the pilot is navigating primarily by reference to features on the ground. It is necessary that he should be able to identify terrain quickly and, in order to assist him, relief is portrayed by contours and shading while only those cultural features are mapped which are important as landmarks or as hazards to aviation. The aeronautical information included can be divided into four sections: first, a selection of symbols associated with airports and providing data about radio frequencies, elevation, runway length, etc.; secondly, technical data about radio navigational aids; thirdly, airspace information relating to controlled zones, prohibited or restricted areas, airport traffic areas, etc.; fourthly, a miscellaneous group which locates obstructions as well as aids to navigation. Instrument Approach and Procedure charts are available for major airports; these are used for radio contact and approaches using instruments in times of poor visibility.

Like that of the marine chart, the design of the aeronautical chart has to suit the conditions in which it will be used. The chart designer has to bear in mind the restricted space available in the aircraft cockpit or flight deck when deciding a maximum sheet size, and any colours he selects for printing his charts must be such that they will be clearly differentiated in the cockpit lighting. This is just one example of cartography meeting a new challenge and producing a specification to cope with it. The story of route maps and charts is full of such challenges as travel has become more extensive, vehicles have increased in speed and range, and completely new modes of travel have come into being. Future increases in extra-terrestrial flight will pose new problems for the cartographer, as will the more extensive use of super-sonic commercial aircraft on conventional global flights.

9 TOWN PLANS AND VIEWS

Modern town plans are primarily clear and functional and convey to the user a considerable amount of information about the urban scene. Rarely, however, are they visually stimulating and rarely do they communicate much feeling of a town's individual character. Indeed the presentation of the complex make-up of a town on a manageably sized sheet of paper has been, and still is, a teasing problem to the most ingenious of mapmakers. There is such a variety of information to convey and so many different user-needs to satisfy. The student of planning wishes to see the layout of streets and buildings, how the land is used, how tall are different buildings; the architect is interested in the style and character of the buildings, their age, the materials of which they are constructed; the student of transport looks for the communications network, the system of one-way streets and the way the traffic flows; the sanitary engineer wants to be shown the system of drainage and sewerage; and so on – not by any means forgetting the short-term visitor or tourist who simply is anxious to be told the easiest way to his destination or where the main features of tourist interest are located. It might seem that a way of satisfying such a diversity of interests would be by using some form of cartographic portrayal based on the technique used in architectural drawings where a series of ground and floor plans together with front, rear and side elevations is provided. Clearly, however, this is impracticable when dealing with the agglomeration of streets and buildings at varying levels which constitutes any town. Compromises are necessary and some aspects have to be jettisoned. If a street layout only is required with an indication of the ground plan of buildings in order that the user may quickly find his way from A to B, then the plan view seen from directly overhead is the answer. It will tell nothing, however, of building heights, architectural styles or building materials used, but it could delineate the transport network and possibly the land use. It won't tell the stranger that the route from A to B, though it looks short on the plan, involves a very steep climb. Generally speaking the visual appearance of a town is communicated to a much less extent than that of the landscape represented by a topographical map. Mapmakers have long been aware of the difficulties and have approached the task of overcoming them in three ways; first, by preparing a prospect of a town as it would be seen in elevation from a low viewpoint; secondly, by raising the viewpoint and constructing a bird's eye view seen in perspective; thirdly, by drawing a plan as it would be viewed from directly above.

From the beginnings of cartography towns have figured prominently on topographical maps because of their administrative, commercial and military significance, but the town plan as an individual entity was a comparatively late development, not being introduced into Britain until the mid sixteenth century.

PLANS AND TOWN VIEWS FOR THE MARINER

Early fifteenth- and sixteenth-century mariners in the Mediterranean region derived valuable information about coasts, ports and harbours from books of islands, or *Isolario*, which contained sailing directions, island maps, harbour plans and town views. The earliest known extant work of this kind, Cristoforo Buondelmonte's *Liber Insularem Archipelagi* (1420), included a crudely colourful portrayal of Constantinople, the bird's-eye-view technique allowing the cartographer to delineate the urban layout clearly along with the containing walls and prominent buildings. The navigator was thus provided with a means of identifying Constantinople as he made his approach and he was also able to find his way around when he had safely made port. A later work, Benedetto Bordone's *Isolario* (1528), contained a splendid perspective view of Venice. Bordone assisted the mariner in several ways: by showing the inlets to the lagoon; by locating the many islands within the lagoon, and realistically depicting their prominent buildings so that they were quickly recognisable from the deck of a ship; by delineating the main canals and by illustrating the outstanding buildings within the city itself. The oblique

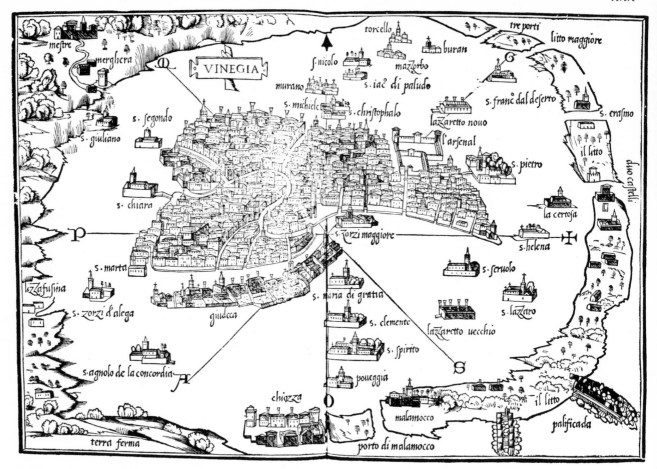

Fig 85 **Bird's eye perspective high-angle view of Venice from Bordone's** *Isolario* **(Venice, 1547 ed). The plan was designed to aid mariners and focuses attention on landmarks and the canal system rather than on the layout of the city.**
Courtesy: **University of Liverpool**

viewpoint used meant that the tightly-packed buildings of Venice obscured the detailed layout of the canal system within the city. Nevertheless, the main arms are clearly visible. Bordone's plan is a useful historical document for a modern researcher who can pick out significant changes; for example, Bordone shows the church of S. Chiara standing alone and reached by a bridge from the main island, but a modern plan shows it located in the heart of Venice's dock system, across the Grand Canal from the main railway station. A comparison with the modern plan will demonstrate the accuracy, for its period, of Bordone's work.

EARLY WOODCUT PLANS

The introduction of the woodcut technique stimulated book illustration and many maps and town views illustrated late fifteenth-century books. Some also featured in broadsheets. The first book to offer a reasonably faithful portrait of a town – a view of Cologne – was Rolevinck's *Fasciculus Tempororum* (1479), but the Nuremberg physician, Hartmann Schedel, provided an extraordinary profusion of woodcut maps, town views and portraits in his *Liber Chronicarum* (1493). In all, 645 illustrations were included and, while it might seem that this amounted to a valuable graphic display of information, the reader must in fact have been baffled by some of the illustrations, for in the economical, though misleading, fashion of the day the same portraits appeared in different parts of the book to represent a considerable number of persons, while town views were also repeated at intervals – in fact sixty-nine different towns were represented by only twenty-two views. Not a great deal of confidence then can be placed in Schedel's

Fig 86 **A dramatic profile of Nuremberg viewed from a low angle. This view from Hartmann Schedel's** *Nuremberg Chronicle* **(1493) gives a good impression of the city's character and includes foreground scenes of local life.**
Courtesy: **University of Liverpool**

portrayal of the urban scene and only a few of the woodcuts can be regarded as even reasonably accurate views. Nonetheless, a lively picture of town life is provided; the prospect of Nuremberg itself is naturally one of the better illustrations as the local craftsmen working on the woodcuts would be familiar with every aspect of the town. They show the reader a handsome walled city of towers, churches, stepped gables and steeply pitched roofs, while outside the walls are located a half-timbered mill, farm buildings, facilities for tilting and heavily-spiked fencing to protect the town gate. The intention of the artist is clearly to convey a recognisable picture of Nuremberg from a well-chosen location and also to tell the reader something of its everyday activities. His low-angle viewpoint, however, reveals nothing of the city's layout and a curiously erroneous impression is given of a

city rising fairly steeply up a hillside and culminating with the great Schloss at the summit. Schedel's view of Genoa is distinctly inferior, being crudely executed and showing little detail, though the overall plan of the port spreading around a mountain amphitheatre and centred on the harbour is recognisable. Even this plan can be regarded as reasonably accurate when placed alongside the view purporting to be of Dover, a portrayal which appears to be little more than a figment of its maker's imagination.

PLANS AND PROSPECTS IN THE COSMOGRAPHIAE

The sixteenth-century German cosmographiae, the most voluminous of which was Sebastian Münster's *Cosmographia* (1544), were copiously illustrated with woodcut maps, views and portraits. Münster provided the great encyclopedia of his day and his work was an important factor in the dissemination of geographical knowledge. *Cosmographia* contained many town plans and views but, like Schedel, Münster misled his readers

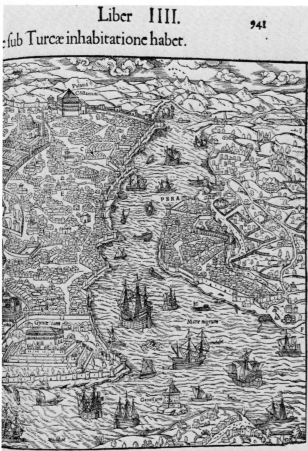

Liber IIII.
sub Turcæinhabitatione habet.   941

locus patriarchatus. D Templum S.Lucæ Euangelistæ. E S.Petrus &c. A Constan
tinopoli ad Perã, quam hodie Gallatë uocant, est traiectus non magnus, quum bombardæ
factus opposītū littus apprehendere possīt, suntcꝫ extra mœnia oppiduli sepulturæ Iudæo-
rum & Turcarum, id quod lapides quadrati erecti indicant. Tormenta ænea quæ uides si-
gnata iuxtamare in littore Perę prope literam F, sunt ea quę Turca à Belgrado, Rhodo &
Buda nostro æuo Christianis abstulit.
                                                     RR ; Anno

Fig 87 **Detail from Sebastian Münster's perspective view of Constantinople from his** *Cosmographiae Universalis* **(Basle, 1550 edition) showing the Golden Horn and the old city with its encircling walls and the suburb of Pera. This is a more refined and detailed view than many by Münster and conveys a remarkable amount of information about the layout of the city and its maritime activities.**
Courtesy: **University of Liverpool**

by repeating the same illustration to portray several different subjects. Indeed it is fair to say that many of the views, crude but lively woodcuts, are hardly remarkable for their realism. Edinburgh, for example, is almost completely divorced from reality and the artist has chosen a low-angle view from the Firth of Forth for his depiction of a city of nondescript architecture situated against an unlikely backdrop of craggy peaks. Münster's bird's eye view of Constantinople is a very different

proposition and it is remarkable that illustrations of such diverse quality could appear in the same volume. The city is viewed from high above so that some impression of street layout can be conveyed. This reveals something of the confusion of streets and alleys but there is a misleading amount of open space due to the artist's exaggeration of street widths. This is an inherent fault of the bird's eye-view technique which seems insoluble. Despite its imperfections, Münster's view communicates a vivid picture of a vitally important city crammed into the peninsula west of the Golden Horn, but already spreading over the water to the suburbs of Pera, Galata and Scutari. The city walls, mosques, palaces, aqueduct and the Leander Tower in the Bosphorus are clearly seen and the harbour is packed with a variety of shipping, for sixteenth-century plan makers did not confine themselves to static objects but attempted to communicate something of the everyday life of each town. A third example from Münster illustrates a variation in technique, for he views the city of Riga in elevation across the Dvina river. Some sense of the urban character emerges from the scene of great churches with slender spires, buildings with stepped gables, town walls, ferries and varied shipping.

HANS LAUTENSACK'S PROSPECT OF NUREMBERG (1552)

The powerful woodcut prospect of the Bavarian city by Lautensack demonstrates the artistry which could be achieved by a fine craftsman. The city is viewed from the south-west from a low angle and the prospect forms an interesting supplement to Schedel's 1493 view. The mill which figures prominently in the extreme right of the latter appears on the left of Lautensack's work, so that we are presented with a new aspect of the city. Lautensack gives a better impression of the topography and his detailed illustration of ecclesiastical and domestic architecture is greatly superior to the earlier work. Nevertheless a comparison suggests that Schedel's depiction of individual buildings was reasonably accurate. In the foreground, outside the town limits, Lautensack provides a vigorous harvesting scene, while a group of gentlemen seated on a knoll point out features of the city, one of them, presumably meant to be the artist himself, sits working at a drawing board. If Münster was sparing of ornament, Lautensack went to the other extreme with an overwhelmingly heavy strapwork cartouche and a flying scroll across the full width of the composition carrying the title. The lettering is in the south German style known as Fraktur, and is well in accord with the generally dark 'colouring' of the whole design.

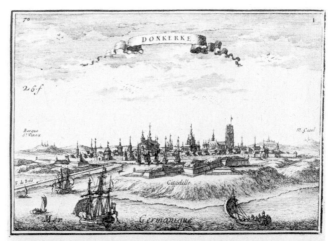

Fig 92 **Blaeu's pictorial plan of the Italian city of Gubbio gives an evocative portrait of the medieval town and is one of many superb views in** *Novum Italiae Theatrum*, **the 1724 edition of Blaeu's town books of Italy published by Pierre Mortier in Amsterdam.**
Courtesy: **University of Liverpool**

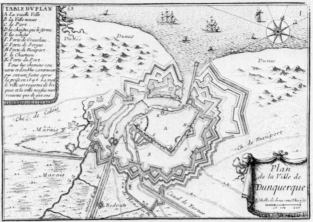

Gubbio is representative of the series. The town is displayed to great effect in its situation at the mouth of a deep gorge; it is girt with defensive ramparts and Blaeu distinguishes no less than forty-five churches and monasteries by a numbering system on the plan which is explained in a reference table. Another table lists prominent buildings, among them the Ducal Palace and the fourteenth-century Palace of the Consuls. A well-preserved Roman theatre stands in the foreground in much the same condition as it is today. In the use of reference tables, the inclusion of much carefully-delineated detail and particularly by the provision of a linear scale of 500 paces, Blaeu shows a new sophistication and an attempt at accuracy which had not been apparent in the work of his predecessors.

MILITARY PLANS

Military strategic planning both for defence and attack often necessitated improved maps and up-to-date surveys. During the seventeenth and eighteenth centuries many urban plans were engraved to serve military needs, often in connection with the elaborate and expensive defensive fortification systems which were introduced by the French military engineer, Sebastian de Vauban. The pages of a contemporary atlas of military plans often featured three-part illustrations: first showing a particular strategic town in elevation; secondly giving a plan view of the town's layout; thirdly a map locating the town within its region. This graphic presentation must have served the military commanders well, for on a single page they could see the approaches to a town, the detailed layout of its defences, and the architecture of its prominent buildings, enabling them to be quickly recognised. In this field French and German mapmakers were specially proficient; the 1696 edition of A. H. Jaillot's *Atlas* edited by Pierre Mortier contains numerous plates of fortifications and plans of sieges while another Frenchman, George Lousise le Rouge, was renowned for his plans of fortifications in the mid eighteenth century. His German contemporary, the Nuremberg engraver J. B. Homann, published many siege plans which provide useful documentary evidence of fortifications and of battle campaigns.

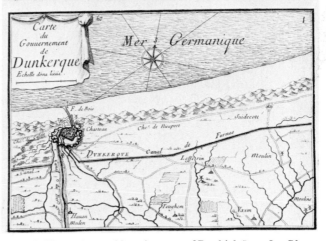

Fig 93 **View, plan and location map of Dunkirk from** *Les Plans et Profils des principales Villes . . . du Comté de Flandre* **by S. de Beaurain (Paris, c.1760)**

TURGOT'S PLAN OF PARIS (1734–9)

Paris, like other great European cities and notably London, Rome and Amsterdam, has been splendidly served by cartographers and a collection of plans devoted to any

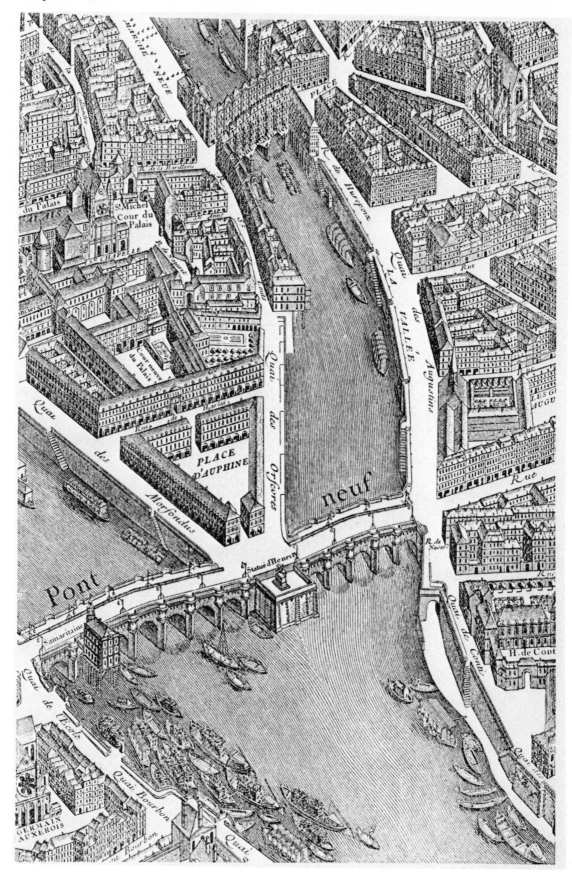

Fig 94 **Detail from Turgot's remarkable plan of Paris drawn and engraved by Louis Brétez and Claude Lucas between 1734 and 1739. The detail shows part of the Ile de la Cité looking east from Pont Neuf.**
Courtesy: **The Librarian, St David's University College, Lampeter**

one of these cities provides a historical record of its growth and also of buildings, streets and squares which have disappeared over the centuries due to decay, redevelopment or wartime destruction. Braun and Hogenberg's 1572 plan of Paris shows many details of the medieval city, the overhead viewpoint allowing its geography to emerge clearly and with reasonable accuracy. Merian's 1615 view is a superb portrait of the city in which the buildings are engraved in considerable architectural detail and a good deal of lively action is revealed in the streets and on the Seine. A 1618 view by the Dutchman, Visscher, is a masterpiece of artistic cartography which conveys the character of Paris to perfection.

Even Visscher's plan pales somewhat when viewed alongside the twenty-sheet perspective plan drawn and engraved by Louis Brétez and Claude Lucas between 1734 and 1739 to a commission by Michel Etienne Turgot. This plan shows the heights of skill and artistry achieved by eighteenth-century designers and engravers in the perspective plan technique. If Turgot's plan is compared with a modern oblique aerial photograph the extraordinary topographical and architectural accuracy of the work stands out. It has great value in recording scenes as they were before being overtaken by urban redevelopment; for example, tall gabled houses are shown, not only crowded within the Isle de la Cité, but also spreading on to the Seine bridges. A notable exception, however, is Henri IV's *Pont Neuf* which was deliberately kept clear of buildings so that the view of the river and the Louvre might be preserved.

By this time, therefore, the perspective view had been developed to perfection and mapmakers were able to present a detailed urban portrayal which provided much information of street layout, disposition of buildings and open spaces, trade and daily life, architectural styles and so on. The plans were highly pictorial in character and would be well worthy to grace the walls of any rich citizen's home. They would also be of interest and use to planners, municipal authorities and students of architecture. They were not, however, the products of a scientifically accurate survey and were not suitable for the taking of measurements in the way that a modern plan might be used. Nor were they quite the sort of thing the visitor to a town would use simply to find his way about – indeed streets were not named though Brétez

used a system of reference numbers on his plan for Turgot.

LONDON AND THE GREAT FIRE

The havoc wreaked by the Great Fire of London in 1666 created an urgent need for new and accurate surveys to help the authorities in their moves to rebuild the city and settle any disputes concerning property boundaries. To this end John Leake was ordered by the City of London authorities to make an 'exact survey of the streets, lanes and churches as they were before the fire'. Leake's scientific survey demarcated the devastated areas, delineated and named burnt-out streets and depicted the ground plan of city churches. He also marked ward boundaries and his work constituted a unique record of the city as it once stood. At a later stage John Ogilby, one of the most interesting and innovatory of British mapmakers, with the support of the authorities, appointed William Leybourne to take charge of an entirely new survey of the city as a whole and of its individual properties. The resulting plan was published in 1677, a year after Ogilby's death, by his step-grandson William Morgan. It was made in twenty-one sheets at the very large scale of one hundred feet to an inch and was the first truly accurate, detailed plan of London. Entirely drawn in plan view, even the buildings, the plan is very modern in appearance with streets named, buildings shaded and public buildings picked out with heavier cross shading. Ogilby and Morgan also prepared a plan at a scale of 300 feet to an inch which covered not only the city itself but also the built-up area outside it.

Among the finest town plans of any period were those of the Huguenot estate surveyor, John Rocque. His plan of London and Westminster (1746) appeared in twenty-four sheets at the remarkably large scale of twenty-six inches to one mile. Rocque portrayed in full plan view a closely-built up city of alleys, courts, inns and churches and his introduction includes an interesting description of the survey methods used:

> The Method followed in making this Survey, has been by ascertaining the Position and Bearings of the Churches and other remarkable Buildings, by Trigonometrical and other Observations from the Tops of Steeples, Towers and other Places, whence such Buildings are visible; by taking the Angles at the Corners of Streets &c. with proper Instruments, and measuring the Distances by the Chain; and by comparing, from Time to Time, the Position of Places, found by this last Method, with the general Observations before-mentioned, so as to correct the one by the other.

The extensive progress towards more scientifically constructed, accurate town plans made during the late seventeenth century, as seen in Ogilby and Morgan's work and that of Rocque, was maintained throughout

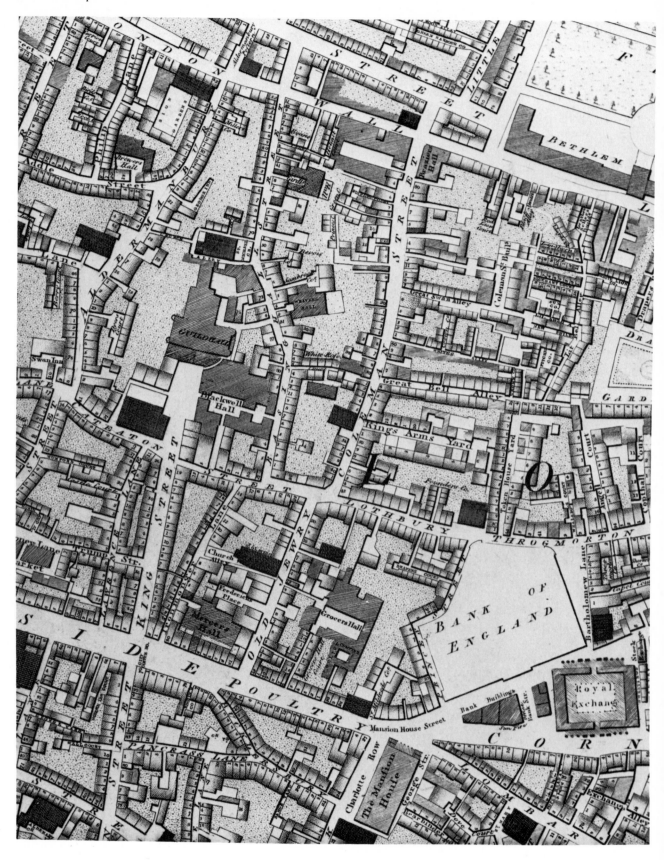

Fig 95 **Detail from Richard Horwood's plan of London (1792–9) at a scale of 26″ to one mile and showing every house, court and alley as well as a great many house numbers.**
Courtesy: **Guildhall Library**

the eighteenth century when provincial towns as well as the capital received the attention of capable surveyors. William Bradford's plan of Birmingham (1754), Isaac Taylor's Hereford (1757) and William Green's classic plan of Manchester and Salford (1794) are examples of the work of the period in England.

The outstanding British town plan of the eighteenth century was Richard Horwood's extremely detailed plan of London (1792–9) produced at twenty-six inches to one mile. Horwood managed to locate every house, court and alley and, although his original intention of indicating every house number was not entirely fulfilled, many numbers are included. In 1807 William Faden issued a second edition of Horwood's plan, adding eight extra sheets to extend the map eastwards so that the major improvements made to the dock system in the early nineteenth century could be recorded. Further editions were issued by Faden in 1813 and 1819 and the plan remained the finest of London until the first edition of the Ordnance Survey twenty-five-inch-to-one-mile series in the 1860s. It has particular value in that the various editions record the extensive changes which took place in the city during a period of rapid growth.

PORTRAYING THE AMERICAN CITIES

The first urban plan to be engraved and published in America was John Bonner's portrayal of Boston (1722) in which he combined the plan view with buildings drawn in profile. In general, however, European map-makers were responsible for the bulk of eighteenth-century plans of American towns. The Frenchman, J. N. Bellin, prepared *Petit Atlas Maritime* (1764) containing detailed plans of maritime towns and harbours which, though small, are of considerable interest, that of Detroit, for example, being the first printed plan of the city. John Rocque compiled *A Set of Plans and Forts in America Reduced from Actual Surveys* (1765), containing thirty plans engraved in his unmistakable style. Rocque's plan of Montreal, for example, conveys a good deal of information which would be valuable for military purposes: the walls surrounded by a dry moat, the Fort which Rocque describes as 'only a Cavalier without a Parapet', the Powder Magazine, 'the Arsenal and Yard for Canoos & Batteaux'.

New York has been the subject of many plans. Of these, a large-scale (one inch to 400 feet) work engraved by Thomas Kitchin and published by Thomas Jefferys

and William Faden in 1776 is outstanding. This was the most accurate plan of New York in the last years of colonial rule and not only provides a good impression of the surrounding geography but also indicates depths in the Hudson and East Rivers. The plan is supplemented by a beautifully-engraved view of the city entitled 'A South West View of the City of New York taken from the Governour's Island' so that the complete work makes an effective combination of plan and profile.

The nineteenth century saw a positive burgeoning of plans, largely functional in character, in atlases and guidebooks and also as single sheet maps. The most distinctive American contribution to urban mapping came with an extensive series of panoramic maps which are unmistakably American in character. Printed by lithography and not usually drawn to scale, these nineteenth-century plans show street patterns with named streets, individual buildings and surrounding landscape. The method of production was laborious; first the street layout was sketched in perspective and then the artist walked the streets and drew in buildings, trees and gardens to suggest a view as seen from some two to three thousand feet. Many of these views were prepared as a consequence of civic pride, being prepared at the instigation of Chambers of Commerce to communicate a visual picture of the attractions of their towns in order to promote their potential as commercial and residential centres. The most ambitious project was Camille N. Dry's *Pictorial St. Louis; The Great Metropolis of the Mississippi Valley* (1875), a mammoth portrayal on 110 plates with useful information about the life of the city printed on the reverse of each plate.

EIGHTEENTH- AND NINETEENTH-CENTURY PANORAMAS OF BRITISH TOWNS

Around the middle of the eighteenth century Samuel and Nathaniel Buck brought a new sophistication and a high degree of accuracy to the town prospect and their work provides an important documentary record of many English towns not long before industrialisation. 'The South West Prospect of Birmingham', for example, shows an unfamiliarly picturesque town, the only hint of possible industrial development being given in the legend which relates that 'by the Industry & Ingenuity of its Inhabitants, & the Advantage of its being an open, free place of Trade . . . it is risen to a Competition with any of the most flourishing Towns in England'. If the Buck views are limited in what they can convey about the urban scene, their usefulness is enormously increased if they are seen as a complement to the contemporary large-scale plans. The Buck panorama of London (1749), a five-plate work in which, unlike many

Plate 21.

Fig 96 **Camille N. Dry's pictorial plan of St Louis was produced on 110 plates to provide a panoramic view of the city in 1875. The complete plan measured nine feet by twenty-four and the detail shows just one of the 110 plates.**
Courtesy: **Library of Congress, Washington**

of its predecessors, accuracy was of prime importance, can be used in conjunction with the 1746 plan by John Rocque, the two together providing a superb survey of the layout, architecture and life of the city.

The introduction of illustrated weekly newspapers during the 1840s provided an unexpected incentive for the publication of large prospects and panoramas which the proprietors were wont to distribute freely as an inducement to prospective subscribers. The *Illustrated London News* issued several between 1843 and the 1870s, while the rival *Pictorial Times* produced a *Grand Panorama of the Thames* fourteen feet in length which had some claim to be the largest engraving in the world. The making of panoramic perspective plans was further rejuvenated by the popularity of ballooning, the basket of a tethered balloon making an ideal viewpoint from which to make detailed sketches of the townscape below. Many fine drawings of British cities were made, including a superb plan of Bristol published by Lazars and a detailed view of London as seen from Hampstead.

The technique, a form of remote sensing, was a forerunner of modern aerial photography from aircraft and space vehicles.

SPECIALISED NINETEENTH-CENTURY PLANS

Many plans were produced in connection with civic and commercial development projects, property transactions and the like. Others met the requirements of transport undertakings, perhaps mapping out the site of a canal basin or the line of a tramway. Particularly useful data about the nineteenth-century layout of British towns is provided by the Tithe Plans produced to accompany the written apportionments of the Tithe Commissioners under the Tithe Act of 1826. When used together, the plans and documents furnish particulars of acreages, ownership and occupiers and give descriptions of pieces of land.

Guidebooks, directories, local histories and gazetteers were published in profusion in Britain and elsewhere during the nineteenth century and often included small maps and town plans. Apart from these, two sets of plans appeared in connection with investigations into local government during the 1830s. The first, by Lt R. K. Dawson, RE, was contained in *Plans of the Cities and Boroughs of England and Wales* (1832), and delineated

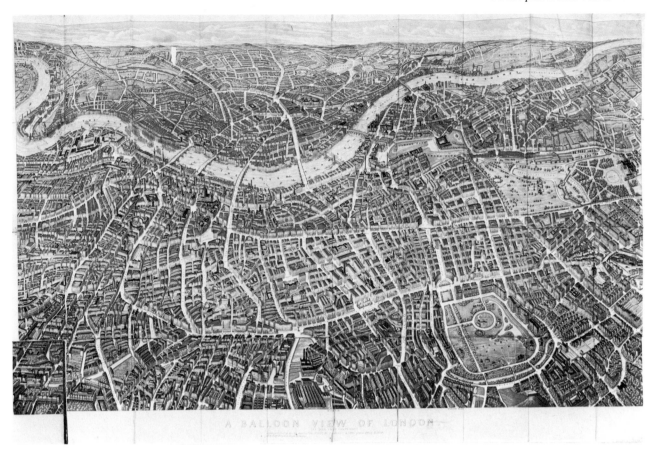

Fig 97 **London as seen from a balloon tethered above Hampstead to the north of the city. Balloon views were popular during the 19th century and this example was published by Banks & Co in 1851.**
Courtesy: **Guildhall Library**

the boundaries of the old boroughs, of proposed boroughs, and the boundaries of parishes established by the Boundaries Act of 11 July 1832. The second set was published in the *Report of the Municipal Corporation Boundaries Commission* (1837) and delineated ward boundaries.

ORDNANCE SURVEY LARGE-SCALE PLANS

Unquestionably the major contribution to British urban mapping in the nineteenth century came from the Ordnance Survey. After 1840 the northern counties of England, together with Scotland, were surveyed at six inches to one mile to provide an excellent urban coverage, but even this comparatively large scale failed to satisfy a substantial body of influential opinion which clamoured for even larger scales. Accordingly, in 1853, a survey of Durham county was authorised at twenty-five inches to one mile and this scale was adopted as basic for Britain,

with the exception of areas of mountain and moorland which did not necessitate such detailed examination. Today the six-inch and twenty-five-inch maps are standard reference tools of local historians and of every local authority. They are not, however, the largest scale plans produced by the Ordnance Survey, for remarkable plans appeared during the nineteenth century at 1:1056 (five feet to one mile), 1:528 (ten feet to one mile) and 1:500 (10.56 feet to one mile). These plans included a staggering amount of information; the ground plans of churches were so detailed as to show each pew and pillar while at street level every lamp standard, drainage gully and pillar box was indicated. The Board of Health plans commissioned by various towns from the Ordnance Survey at 1:500 are worthy of special mention on account of their quality and their function. The Public Health Act of 1848 empowered the Board of Health in London to send Superintending Inspectors to report on the sanitary conditions in particular towns. As a result of these surveys it was sometimes necessary to instal properly designed sewerage systems and ensure adequate piped water supplies. To this end plans were commissioned at the ultra-large scale of 1:500 to provide an accurately-detailed basis for planning sewerage schemes. One of the finest of these plans, that of Warwick, has been

Fig 98 **Detail from Ordnance Survey plan at scale of 1:528 (10 feet to one mile) showing the centre of the Cheshire market town of Nantwich (1878).**

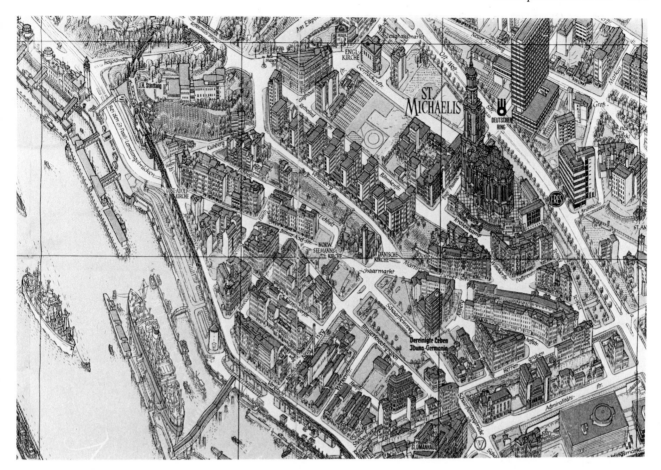

Fig 99 **Detail from the superb modern pictorial plan of Hamburg by Hermann Bollmann with highly detailed drawing of architecture and shipping.**
Courtesy: **Bollmann Bildkarten Verlag, Braunschweig**

recently reproduced in facsimile and apart from being a most beautiful plan it displays various new features. The many items indicated by abbreviations include those which are of particular importance to public health, i.e. sewer gates, privies, piggeries, dung pits, water houses and cisterns along with those of more general interest such as barns, summer houses, public houses, coach houses, etc. The height of each property above sea level is indicated in tiny figures in order that new sewers could be planned to have a proper fall. Visually the most exciting aspect of these plans is the beautiful depiction of parkland, woods and gardens, the trees being drawn from a slightly oblique angle in contrast to the plan treatment of the remainder.

TOWN PLANS TODAY

Street plans proliferate in our modern society and are available in several guises: as folded sheets, as street atlases, sometimes in colour, sometimes in black and white only. Almost invariably they are supplemented by a gazetteer of street names. The town prospect has disappeared and the modern plan is generally a purely functional aid to route finding and locating one's position.

The only way to provide a satisfactory composite view of all the aspects of a modern city is in atlas form with individual plans devoted to specific themes. Berlin, London and Paris have monumental volumes which thoroughly analyse each aspect of the urban environment. Smaller cities too have been favoured with the atlas treatment including Durham, Harrogate and the outstanding *Atlas of Portsmouth*.

No survey of urban mapping is complete without a mention of the fascinating perspective plans or 'bildkarten' produced by Bollmann Bildkarten of Braunschweig. Originally conceived by the late Hermann Bollmann as historic documents which would provide a visual record of the rebuilding of devastated German cities at various stages after World War II, the plans convey such a vivid impression of urban character that they never fail to arouse interest and excitement. The plans are based on what Herr Bollmann described as an 'optical perspective'

Fig 100 **Bollmann's most ambitious project was the view of Central Manhattan from which this detail is taken. In order to achieve clarity of presentation it has been necessary to exaggerate the street widths.**
Courtesy: **Bollmann Bildkarten Verlag, Braunschweig**

in which the scale does not diminish with distance. The viewpoint is that of an observer looking into the town from an oblique angle and each building is drawn with a considerable amount of architectural detail. In his plans of ports such as Hamburg, Bollmann included a great deal of activity in the rivers and harbours but, curiously, his streets and railways are empty. The Bollmann firm has produced plans of many German towns and cities as well as an atlas of village plans in the Peine district near Hannover. Activities have now been extended to foreign towns and perhaps the most remarkable plan of all is that of central Manhattan, though it is arguable whether such a city is as suited to the Bollmann treatment as the historic towns of Europe.

If it remains impossible to capture every facet of a city in a single plan then surely Hermann Bollmann and his successors have come as near as is humanly possible to attaining that objective.

10 THEMATIC MAPS

The topographic map may be regarded as a general-purpose map, one which is constructed from ground, air or even space survey and which serves as a source from which a variety of information can be extracted. In sharp contrast is the thematic or special-purpose map, a blanket term covering an infinite number of topics, for it is true to say that more or less any theme can be presented in map form providing that suitable data are available. The thematic map essentially concentrates on a single theme in the environment – physical, social or economic – or at most two or three themes which are in some relation to each other. Thematic maps basically consist of a framework of selective topographical data – coastlines, rivers, administrative boundaries, placenames and relief – which serves to provide locational information and often to add meaning to the thematic information being mapped and which is derived from a specialised survey or from available statistics.

Thematic mapmaking is a comparatively recent development, largely a product of the seventeenth, eighteenth and nineteenth centuries when man began systematically to examine his relationship with the environment. The nineteenth century, in particular, saw a great burgeoning of thematic mapmaking which could largely be attributed to the proliferation of new data derived from census taking and the extensive recording of statistical information about multifarious phenomena. A further contributory factor lay in the virtual completion of the geographical world map; as the task of locating places on a world scale drew to a close mapmakers began to consider not just 'where is it?' but 'what is there?' and eventually 'how much is there and what is its value?'

Today the making of thematic maps occupies a major part of the cartographic spectrum and makes a significant contribution in various sectors – education, administration, military, industry and trade, transport, leisure activities and planning.

Before considering the historical development of thematic maps it may be appropriate to define the main classes of the *genre*. The first category is termed the *qualitative* map, i.e. one which features distributions without any element of value, amount, percentage or ratio, merely indicating the kind of phenomena occurring in specified locations. Such maps have been described as 'either/or' maps, implying that a place or an area is either one thing or another – either pasture or arable for example. Qualitative maps were the only type of thematic map produced up to the late sixteenth century. The second category is the *quantitative map* which, as its name suggests, adds information about amount or value to the simple indication of phenomena.

Each main category may be sub-divided into three, for both quantitative and qualitative data may relate to point, line or area situations. It is easy to call to mind situations in which the phenomena occur at a series of point locations, e.g. in qualitative terms all the theatres in a large city such as London could be plotted on a map; this could be developed into a quantitative situation by obtaining data about the average nightly attendances over a fixed period for each theatre, then using an appropriate statistical presentation technique to display the quantitative information in map form. A map which simply delineates a communications network, e.g. the routes operated by British Rail or the Greyhound Bus Company in the United States, is an instance of a qualitative linear map. A quantitative element could be added by using a technique called 'flow-line' to demonstrate how many passengers or what tonnage of freight is carried on each section of line. The land-use map is one example of a qualitative area map – geology, vegetation, soils are others. There are many instances of quantitative area maps but one which shows the varying densities of population over certain areas will serve to indicate the nature of this quite common type of thematic map.

Quantitative maps may be used to map either *absolute* or *derived* data; the former can be exemplified by referring to a map which indicates absolute heights above a common reference level or *datum*; a map using derived quantities would be concerned with averages, ratios,

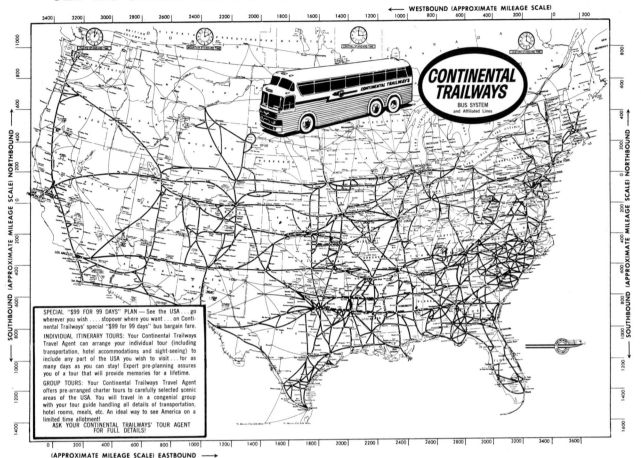

Fig 101 **A simple use of a qualitative map based on linear data used for publicity purposes by the Continental Trailways Bus System.**
Courtesy: **Continental Trailways**

percentages or densities – in other words, cases where some manipulation of the statistics is necessary.

THE ANCIENT WORLD

In order to administer effectively the great empires of the ancient world two main types of thematic or special-purpose map were required; first, the vital route map which facilitated the travels of merchants, military personnel and administrators; second, the cadastral map which has been important throughout mapmaking history, for it is basically one which delineates property boundaries and usually adds information about their location, area and value for fiscal purposes.

The classification of cadastral maps as thematic illustrates the difficulties of drawing a hard and fast line between topographic and thematic maps for it might be argued that cadastral maps are in essence topographical. On the other hand, they are made for a specific purpose and their topographical detail is usually supplemented by statistical information.

The earliest cadastral maps are believed to have originated in Babylon and Egypt, the earliest known extant specimen being a clay tablet plan from central Mesopotamia which dates from *c.*2200 BC. Though no maps have survived from ancient Egypt it seems certain that the Egyptians would be forced to prepare frequent plans showing property boundary delineation because of the Nile's propensity to overflow its banks and wash away boundary markers. The Romans made extensive use of cadastral maps prepared by a land surveyor or *agrimensor* using a technique called *centuriation* in which domains were divided into 'centuries', usually rectangular plots, which were of equal size on any one site. The purpose of this land division was generally that of settlement in domains acquired by conquest. Veteran legionnaires were often allotted holdings in this way. Substantial fragments of a Roman cadastral map on stone have been found at Orange in Provence along with an inscription recording that the Emperor Vespasian ordered the making of the map with records of the annual rental from each 'century'. The fragments show rivers and roads within 'century' boundaries together with owners' names, holdings and so on.

Hundreds of years later an important cadastral survey was made in France by the Intendant of the Duchy of Savoy in 1728–38. Each land parcel in the survey was numbered in accordance with a register giving the name and status of its owner, the type and quality of the land and its area. With the aid of such detailed village plans an illuminating picture of eighteenth-century village life in Savoy can be seen. Such surveys were followed by a national cadastral map of France ordered by Napoleon I in 1807. A survey of English and Welsh parishes under the Tithe Commutation Act took place from 1836 to 1851. In the late eighteenth century much cadastral mapping was undertaken in the United States following the opening up of the Public Domain of the United States for private ownership. The maps produced became an important adjunct of both urban and rural land management.

Route maps are discussed in some detail in another chapter but it is relevant to state here that the earliest surviving road map was discovered in Syria and was made to illustrate a Roman road from Byzantium to the Danube Delta. Drawn on parchment, it dates from *c.* AD 200–50.

Further types of map which might be classified as 'thematic' or 'special-purpose' are discussed in other chapters. They include navigational charts and the extensive class of theological maps produced during the dark ages of cartography.

MAPPING THE SEA BED

For several centuries there were no radical innovations in the field of special-purpose mapping and it was only during the Renaissance that new avenues began to be investigated. In 1579, Abraham Ortelius published an augmented edition of *Theatrum Orbis Terrarum* in which the geographical section was supplemented by a series of historical maps entitled *Parergon* (or 'by-work'). Despite the new theme, the cartography remained qualitative in character and it was not until 1584 that any form of quantitative technique was utilised. In that year a Dutch surveyor, Pieter Bruinss, prepared a manuscript chart of the Spaarne river on which

he drew lines symbolising imaginary lines joining points on the river bed which were at an equal vertical distance below the surface. Bruinss is thought to be the first person to employ such a device and his discovery was significant for, despite a long interval before a similar technique was applied to land maps, he had pioneered a method which could be used to solve the long-standing problem of producing a quantitative picture of relief. Bruinss' innovatory technique of drawing in lines along which values are constant has been given the generic term of the *isoline* technique. The particular type of line he used, which related to points of equal depth below a given datum is termed an *isobath*.

Despite its subsequent importance, Bruinss' work evoked little response from his contemporaries, and it was not until a century later, in 1697, that the next known use of isolines appeared. This took the form of a manuscript map by the Dutch mapmaker, Pierre Ancelin, and depicted part of Rotterdam and the bed of the River Maas. A further use of isolines, yet again in the form of isobaths, came in 1715 when an English naval officer, Nathaniel Blackmore, produced a sea chart of the coasts of Nova Scotia. To this date all experiments with isobaths were confined to river beds or coastal waters and the first known attempt to delineate isobaths further from the shore (and also the first printed survey to feature the technique) was a chart of the *Golfe du Lion* which appeared in *Histoire physique de la mer* (1725) by the Italian count L. F. Marsigli. Four years later the Dutch engineer, Samuel Cruquius, published an engraved map of the bed of the Merwede River, a minor waterway of no great depth, using the isobath technique.

PHILIPPE BUACHE, A PIONEER THEMATIC MAPMAKER

The work already mentioned contributed in no small measure to the evolution of the isoline and to thematic mapmaking but was generally intended for practical use as an aid to navigation. The work of the Frenchman, Philippe Buache, was on a rather different level, for he exploited techniques of thematic mapmaking to further his theories about the structure of the earth. In 1729 Buache was appointed *premier géographe du Roi* and a year later became the first geographer member of the Academy of Sciences. Many of his theories are expounded in the Academy's transactions and one of his major studies was presented to the Academy in 1752 with the title 'An Essay of Physical Geography, wherein it is proposed to present General Views on what may be called the Framework of the Globe, composed of mountain systems that cross seas as well as continents; with

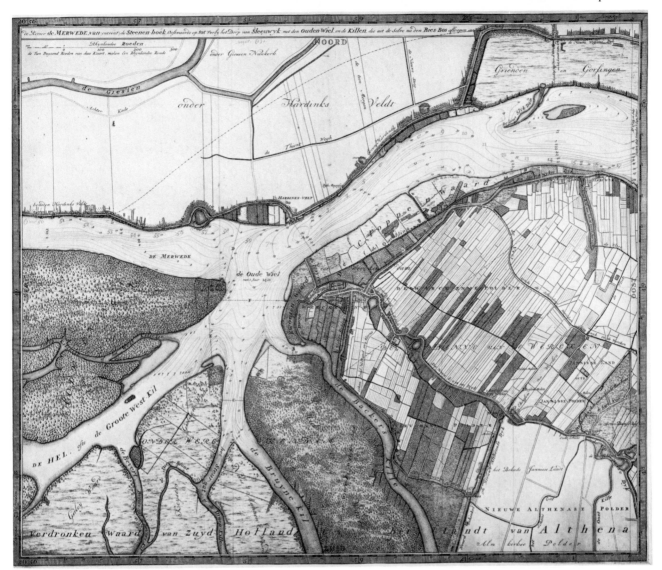

Fig 102 **Nicholas Samuel Cruquius, a Dutch engineer, pioneered the use of** *isobaths* **in his splendid map of the Merwede river (1729). Apart from its interest as an innovatory quantitative map there is a particularly beautiful indication of land use and topographical detail.**
Courtesy: **British Library Board   30965 (1)**

some particular remarks on the different basins of the sea, and on its interior configuration'. This essay was illustrated by a remarkable map of the world, drawn on a north polar projection, on which Buache demarcated the major global mountain systems, drainage basins and undersea topography, in an attempt to demonstrate his view of the entire mountain system of the globe as an inter-connecting whole. Buache then sought to strengthen his argument by presenting another map, this time a bathymetric map of the English Channel, through which

he illustrated his argument that there was an underwater link between the surface features of the terrain on both sides of the Channel. His chart took in the area from the Scillies to the Rhine–Maas estuary and he mapped the configuration of the sea bed by drawing in isobaths at ten-fathom intervals. Above the map was a graphic profile of the sea bed at the centre of the Channel which, when viewed along with the map, must have presented an exciting picture of the unseen Channel depths to the Academy members.

In addition to his use of isobaths, Buache was occupied with a further series of thematic maps in collaboration with a naturalist, Jean-Etienne Guettard, who had an absorbing interest in the distribution of metals and minerals. Guettard's researches served as the basis of three mineralogical maps by Buache – of Switzerland, of the Middle East and of north-eastern

Fig 103 **Isogonic map (1701) by Edmond Halley entitled** *A New and Correct Chart shewing the Variation of the Compass in the Western and Southern Oceans*. **This was the first published map to employ isolines.**
Courtesy: **British Library Board   977(4)**

America. The maps appeared in the Academy Mémoires of 1751–2 and each featured a lengthy 'Explanation des Caractères' which provided the key to the symbols Buache devised to represent a variety of minerals.

In the general history of mapmaking Buache perhaps occupies a minor role but his contributions to thematic mapmaking have been of great significance, notwithstanding the fact that his theories were ridiculed for a long time after his death.

THE WORK OF EDMOND HALLEY (1656–1742)

The English astronomer, Halley, Fellow of the Royal Society, also earned an important place in the history of thematic cartography. His genius was many-sided, extending from the publication of one of the earliest life tables for insurance purposes to the work on comet orbits which enabled him to predict the return of the comet named after him. Halley's mapmaking activities largely stemmed from two years spent in taking celestial observations on St Helena and a voyage in 1698–1700 as commander of a small vessel, the *Paramour*, the first instance of a sea voyage made solely for the purpose of taking scientific observations.

Halley produced his first thematic map as a direct result of his sojourn in St Helena. It took the form of a star chart of the constellations of the Southern Hemisphere. In 1686 he published a far more significant map, a terrestrial monothematic map concentrating on the direction of prevailing winds in latitudes 33°S to 33°N. Halley took care to choose an appropriate projection, the Mercator, for what is generally looked upon as the first meteorological chart, and superimposed a grid of latitude lines at 10° intervals and longitude at every 15°. Halley's chart was plain in the extreme, for in order to concentrate the observer's attention on the main theme he provided virtually no information on the land areas, apart from a few locational names and a hazy delineation of rivers. He used a system of tapering lines to indicate the general directions of the winds and only utilised the much more graphic arrow south of the Cape Verde Islands in an area labelled 'Calms and Tornados'.

Halley's voyage in the *Paramour* was sponsored by the Royal Society specifically to investigate the earth's magnetic field. On the basis of observations taken on the voyage Halley prepared *A New and Correct Chart shewing the Variation of the Compass in the Western and Southern Oceans* (1701), a map which is a landmark in the lineage of the isoline, for on it Halley used, for the first time on any printed map, lines of equal magnetic declination or *isogones*. It is believed that in this connection Halley may have been influenced by the work of an

Italian, Christoforo Borri, whose map has disappeared but is referred to in a book published in 1641 to which Halley had access. Halley's isogonic chart is far superior in its execution to his earlier meteorological chart and some concession is made to decoration in the form of three ornamental cartouches. In 1702 Halley issued a second isogonic map in which data provided by other scientists was employed to extend the area of his original map to cover the Indian Ocean. Halley is also credited with publishing the first scientific tidal chart, although a diagrammatic map of tides had been prepared by Guillaume Brouscon in 1545. Halley's chart was based on observations made in the English Channel between June and October 1701. As well as the standard information about coasts, hazards and soundings provided in conventional marine charts, Halley showed the tidal situation by means of Roman figures indicating the times of high water and arrows showing the direction of tidal currents. Halley's chart was much copied and there were no further developments in charting tides until William Whewell made a tidal chart of the world and one of the British Isles in 1835. In 1848 and 1851 F. W. Beechey prepared maps of the tidal streams of the Irish Sea and the English Channel, and of the North Sea in which he extended the scope of the tidal chart by showing lines of direction of stream, lines of equal range, and direction of turn.

PHYSIOGRAPHIC MAPPING

Although Thomas Digges made reference to the heights of fourteen points above sea-level in his 1581 plan of Dover Harbour it was Christopher Packe, an Englishman, who was the first to show true spot heights in order to represent the positions of measured altitudes above a prescribed datum. Packe, a medical practitioner, published a *Specimen of a New Philosophico-Chorographical Chart of East Kent* to cover the countryside within a three-mile radius of Canterbury. Seven years later he extended the map to cover the whole of East Kent at a scale of approximately one and a half inches to one mile in a pioneering attempt to display the physiography of a sizeable tract of land. Packe's spot heights were measured by barometer from a datum of low-water mark in Sandwich Bay. Packe also used heavy shading of slopes in his depiction of drainage and relief and his map was also one of the earliest English maps to distinguish between types of soil.

In Ireland Charles Baylie and John Mooney included some 160 spot heights in a survey of the Demesne of Carton, County Kildare (1744). An officer of the Corps of Engineers, Milet de Mureau, expounded the principle of spot height representation in a *Mémoire pour faciliter*

Fig 104 **Christopher Packe's attempt to show absolute heights by converting barometric readings to elevation figures in his** *A New Philosophico-Chorographical Chart of East Kent* **(1743). Packe used heavy hatching in a semi-pictorial impression of relief and colours to indicate soil types. He also used symbols to show such features as beaches, chalk pits and gravel pits.** *Courtesy*: **British Library Board   PS 903519**

les moyens de projeter dans les pays de montagnes (1749), but the practice of featuring spot heights on French maps was more widely adopted during the 1770s and 1780s, mainly by the Corps of Engineers.

The introduction of spot heights had introduced a quantitative element into physiographic mapping but, despite the earlier use of submarine contours, it was not until 1777 and 1782 respectively that proposals were made by Meusnier and du Carla that land surface configuration should be represented by lines joining points of equal elevation, i.e. contour lines. In 1791 a French engineer, J. L. Dupain-Triel, published a map of France using crude contouring, later adding tints between contours. He is generally acknowledged to be the first to use a system of contour lines over a wide extent of terrain.

In France, experiments with contours were continued and by 1810 engineers had begun a series of detailed contour maps and firmly established contouring as a definitive method of terrain depiction. Britain continued to lag behind and it was not until 1839 that contours were introduced on to the sheets of the 1:10,560 survey of Ireland, and a further fourteen years elapsed before they were adopted in England.

In 1806 the distinguished German geographer, Karl Ritter, followed Dupain-Triel's example in the use of hypsometric tints, using graded bands of grey between contours, the intensity decreasing as elevation increased. The convention used today in black-and-white mapping is the opposite of Ritter's in that it is normal to increase the intensity of tone with increased altitude.

THE ISOTHERMAL MAP OF ALEXANDER VON HUMBOLDT (1769–1859)

Von Humboldt, like Ritter, one of the most distinguished names in the history of geography and, also, like Ritter, associated with the German publishing house of Justus Perthes in Gotha, undertook research in several areas including mineralogy, physics, oceanography and climatology, but it was his work in connection with the last-named that earns him his distinctive place in cartographic history. In 1817 he presented a paper to the French Academy of Sciences on the subject of *isotherms* or, to use his definition, 'curves drawn through the points on the globe which receive an equal quantity of heat'. In the same year he published a map, or rather a diagram in that it included no coastlines or other topographical detail, entitled *Carte des lignes Isothermes par M.A. de Humboldt*. In preparing this map he was mainly concerned with the relationships between isotherms and geographical latitudes as he aimed to demonstrate visually his findings that temperatures on the western sides of continents were markedly milder than those of the east, a theory which was at variance with the old classical notion of climatic zones based on latitudes. Humboldt marked on his map eight latitude lines at 10° intervals from 0° to 70°N and plotted the average temperatures at thirteen locations in America, Europe and Asia. Isotherms at 5° intervals show lows at concave parts of the curving lines at 80°W (eastern United States) and 116°E (eastern China) and high points at the convexity of the curves at 8°E (Western Europe and West Africa). Humboldt was also interested in relationships between temperature and altitude and in this context added a smaller diagram below the main one to demonstrate the effect of altitude on temperatures at latitudes ranging from 0° to 60°. Despite its cartographic simplicity, Humboldt's map demonstrated the efficacy of the isoline as an analytical tool and a basic communication technique.

Kosmos was the summation of Humboldt's life work. Two of its four volumes have content of a general nature concerning the physical sciences and their inter-relationships; the third volume is devoted to astronomy and the fourth to physical geography. An accompanying atlas to the last two volumes ranges over the field of thematic geography. Humboldt devised a four-fold classification of rocks into volcanic, sedimentary, metamorphic and conglomerate, used lines of mean annual temperature (*isotherms*) in his climatological work, and examined the distribution of flora and fauna.

THE 'PHYSIKALISCHER ATLAS' OF HEINRICH BERGHAUS

Another outstanding nineteenth-century geographer/ cartographer to be connected with the Justus Perthes Cartographic Institute was Heinrich Berghaus, a man who produced many important maps but whose renown stems from his *Physikalischer Atlas* (Gotha, 1842), a remarkable work which was the earliest atlas to be composed of thematic maps and one which graphically synthesised early-nineteenth-century researches in a number of disciplines. The eight sections cover meteorology, hydrography, geology, magnetism, botany, zoology, anthropology and ethnography. It is especially noteworthy in its development of the isoline technique to portray varied meteorological data, for Berghaus not only employs *isotherms*, *isohyets* (lines of average annual precipitation) and *isobars* (lines of equal barometric pressure) but even *isobronts* (lines of equal frequency of the thunderstorms). Berghaus's maps are effective communication devices due to their good design and layout combined with aesthetically pleasing hand-colouring. It is interesting to note that the colouring was mainly undertaken by women and children in their own homes. Although colour printing by the newly-developed process of lithography was available, Berghaus's publisher dismissed it as too expensive, so that the maps had to be printed from the earlier process of copper engraved sheets. Berghaus managed, however, to introduce interesting tonal effects on his world map of precipitation by means of aquatint etching – a good example of a cartographer using unconventional methods of execution to secure just the effect he required.

THE THEMATIC MAPS OF ALEXANDER KEITH JOHNSTON (1804–71)

On no-one did Berghaus exercise a greater influence than on the Edinburgh engraver, Alexander Keith Johnston, who in 1826 joined his brother William in the publishing business he had founded a year earlier and which has since earned world fame as W. & A. K. Johnston. Keith Johnston travelled in Austria and Germany

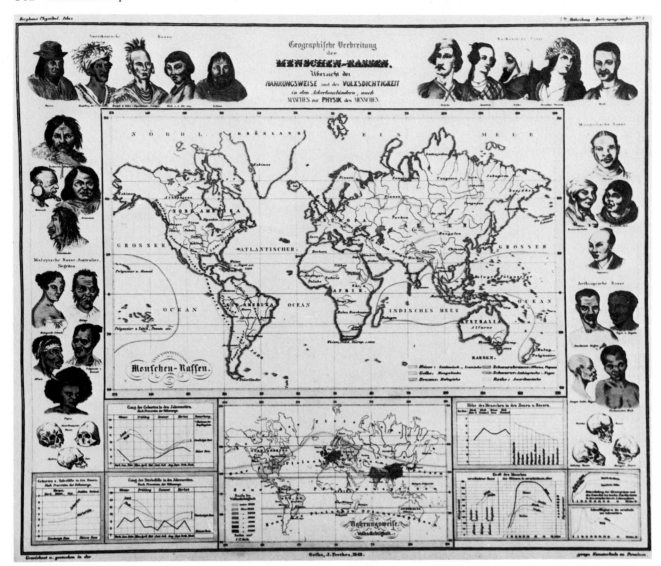

Fig 105 **Ethnographic map from Heinrich Berghaus's** *Physikalischer Atlas* **(1850) with illustrations of different ethnic types.**
Courtesy: **University of Liverpool**

in 1842 and came to an arrangement with Berghaus resulting in the publication of four of the latter's maps in Johnston's *National Atlas of historical, commercial and political geography* (1843). The four maps illustrated the spatial distribution of food plants, the distribution of air currents, the disposition of mountain chains in Europe and Asia, and the isothermal lines of Humboldt. An edition of this atlas published in 1849 is regarded as being the first general atlas to be colour printed by lithography. Johnston's association with Berghaus resulted in a further work, *The Physical Atlas . . . illustrating the Geographical Distribution of Natural*

Phenomena . . . based on the Physikalischer Atlas of Professor H. Berghaus with thirty maps covering four main themes – geology, hydrology, meteorology and natural history. A second edition of the atlas (1856) included a map illustrating the distribution of diseases. Keith Johnston was appointed as 'Geographer to the Queen in Scotland' in recognition of his services to cartography and geography.

THE MOVEMENT IN THEMATIC CARTOGRAPHY TOWARDS SOCIAL AND ECONOMIC GEOGRAPHY

In the mid nineteenth century the attention of thematic mapmakers showed a marked shift towards social and economic themes and a variety of new techniques were introduced to permit the visual communication of a wide range of phenomena.

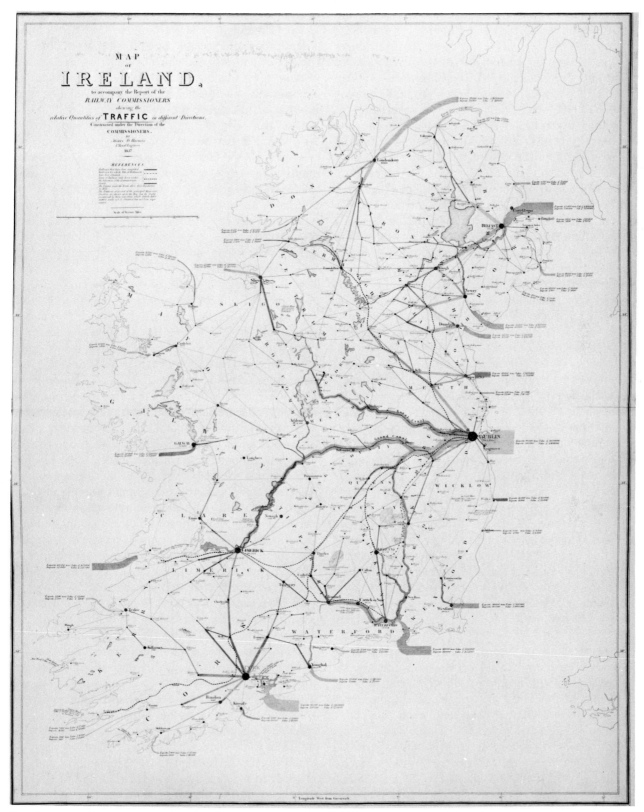

Fig 106 One of Henry D. Harness's three maps of Ireland (1837) in which he used innovatory techniques of statistical presentation; in this map flow lines to show relative volume of traffic and graduated circles for urban populations.
Courtesy: British Library Board   Maps **145.e.29**

Among the most significant thematic maps of the mid century period were three which formed part of an *Atlas to accompany the Second Report of the Commissioners appointed to consider and recommend a General System of Railways for Ireland* (1838). The maps were compiled by Henry D. Harness, an army lieutenant in the employ of the Irish Railways Commission and were aimed to communicate graphically to the Commission information concerning rural and urban populations in Ireland and about the flow of traffic by public conveyance. Harness devised highly successful graphical presentation techniques to illustrate these themes – indeed they were so effective that each of Harness's techniques has become an essential part of the thematic mapmaker's stock-in-trade. In his first map he introduced the *flow-line technique* in which the width of a band drawn along a routeway is made proportional to the amount of freight or the number of passengers carried. Harness in his population map used four grades of aquatint shading to represent the densities of rural population, combined with graduated circles for urban populations, the size of the circles being proportional to the number of persons. Harness also pioneered the *dasymetric technique* in his indication of rural population, for he did not confine his graded shadings by administrative boundaries with their unrealistic limitations, but used the refined method of shading homogeneous areas. Harness's third map was another flow-line map, this time showing the number of passengers carried by public conveyance – the first map had used the technique to indicate the amount of goods carried.

Harness thus pioneered flow-lines, graduated or proportional circles, graded shadings to represent densities, and the dasymetric technique. It is considered possible that these innovations may have been inspired by Captain Thomas Larcom, superintendent of the Ordnance Survey office in Dublin at the time but, whatever the source of Harness's inspiration, he had achieved a remarkable advance in statistical presentation techniques.

By 1840 most of the carto-statistical presentation methods in use today had been developed and mapmakers could cope with quantitative distributions in point, line and area situations; the isoline concept was firmly established and Harness had employed flow-lines, density shadings and graduated symbols. A further technique was introduced about this time, and rather surprisingly it was medical men rather than cartographers who were responsible. In 1849 Dr Thomas Shapter used dots, crosses and circles of uniform size to locate deaths from cholera in Exeter in 1832, 1833 and 1834. A better known map is that made by Dr John Snow in 1855 to illustrate deaths from the cholera epidemic in the Broad Street and Golden Square districts of London in 1854. Snow plotted two distributions – deaths from the disease, by uniformly-sized dots, and the site of pumps from which drinking water had been obtained, by small crosses. His aim was to ascertain whether any relationship could be established between the two factors and he was able to connect the cause of the outbreak with contaminated water from one particular pump in Broad Street.

The renowned German geographer, Augustus Petermann, who worked in London between 1847 and 1854, made effective use of statistical cartography in the course of his researches into the correlation between population density and the incidence of disease. In his *Cholera Map of the British Isles* (1852) Petermann symbolised the relative mortality from the disease over different areas by means of shading patterns and used a similar technique in *A Map of the British Isles, elucidating the distribution of the population based on the 1841 Census* and in *Scotland, Distribution of the Population. Census of 1851*, following the example of Harness in employing graded shadings for population density allied to graduated circles for urban populations. In 1851 Petermann explored new ground in a map based on the 1851 Census which 'showed the places where certain well-defined works and manufactures are concentrated' by means of point symbols. This is considered to be the first distribution map exclusively concerned with the pinpointing of industrial concentrations, and in so doing it conveys a useful picture of the new geography of population and industry consequent upon the development of the British railway system. Petermann also saw the great cartographic possibilities of the United States Census data and used a variety of presentation techniques to present visually the Census information in maps which accompanied articles in his periodical, *Petermanns Geographischen Mittheilungen*, 1856–9.

In 1846 and 1855 a Dane, Niels Ravn, pioneered the use of *isopleths* in representing population density and this meant that by the middle of the century all the standard techniques of statistical presentation used by today's cartographers had been devised. Statistical cartography had achieved international recognition, particularly in Europe where a French engineer, Charles Joseph Minard, published important transportation maps in 1845 showing the volume of traffic carried on railways, canals and roads in a given area. Minard did much to popularise the flow-line technique and also introduced a refinement to the proportional circle by dividing it into sectors to produce the now-familiar *pie graph*, a device in which the area of the circle is made to represent a total figure, a population for example, and the segments are made to represent the percentages of

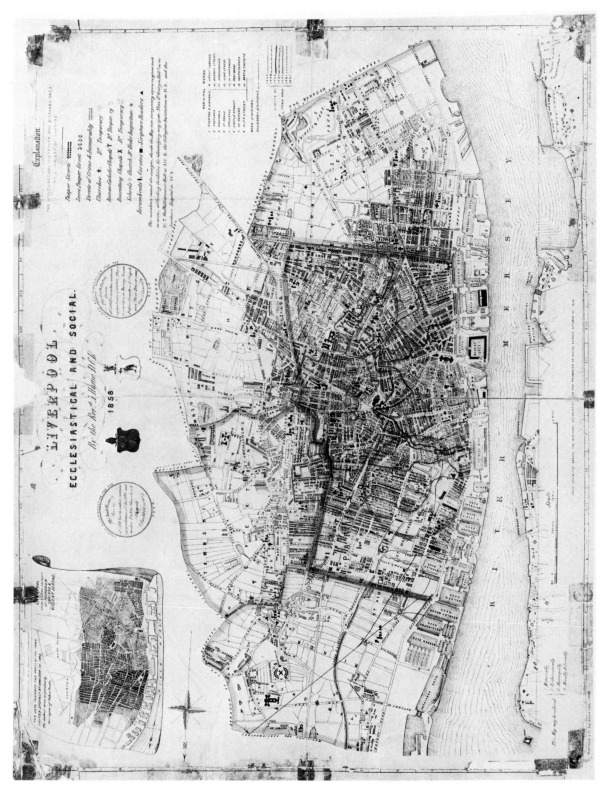

Fig 107 **The Rev. A. Hume's map of social and ecclesiastical conditions in Liverpool (1858). Hume distinguished Pauper Streets, Semi-Pauper Streets and Streets of Immorality and his inset map shows parts of the town affected by cholera in 1849.**
Courtesy: **University of Liverpool**

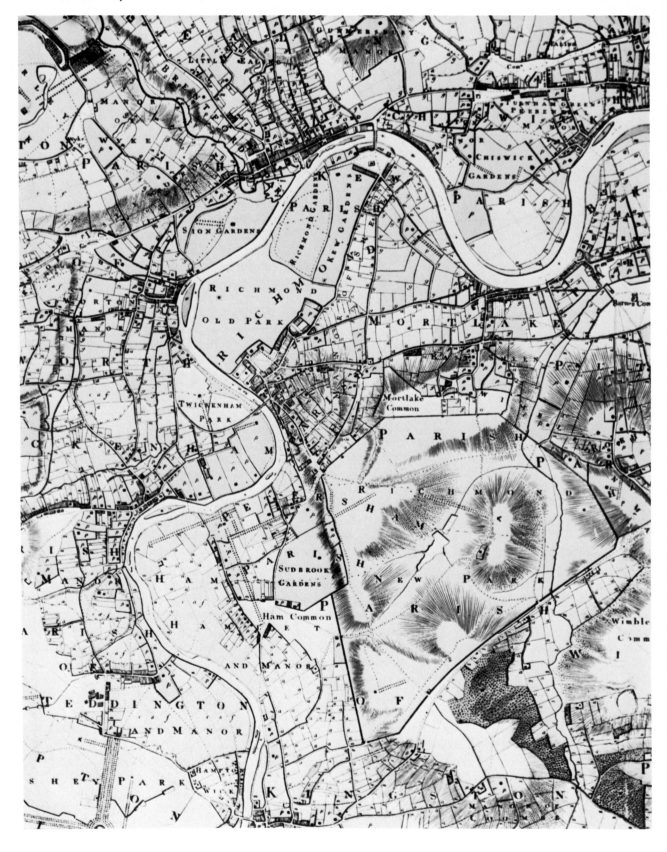

Fig 108 **Detail from Thomas Milne's land utilisation map of the London region (1800). Milne used key letters to distinguish seventeen types of land use. Reproduced from a facsimile by the London Topographical Society.**
Courtesy: **The British Library and the London Topographical Society**

the total figure occupied by certain groups within it, perhaps religious or ethnic groups.

Social conditions occupied the attention of several researchers. The Rev. A. Hume, for example, prepared a map showing the religious state of England (1860) and an interesting plan, *Liverpool Ecclesiastical and Social* (1858), in which he delineated not only conditions of poverty but areas where crime and immorality were widespread. In somewhat similar vein Charles Booth published a series of 'poverty maps', in a graphic display of the results of his investigations into the conditions of London's less affluent citizens, using seven colours to indicate degrees of poverty. In Dublin, surveyors working for the Ordnance Survey of Ireland included several maps of social conditions in *A Report of the Commissioners appointed to take the Census of Ireland for the year 1841*. Four of these maps used the differential shading method to show density of population, classes of housing, literacy and livestock. The remaining illustration was a plan of Dublin coloured to differentiate classes of residential and shopping districts.

STATISTICAL CARTOGRAPHY IN THE UNITED STATES DURING THE NINETEENTH CENTURY

Statistical mapmaking in the United States was slow to gather momentum but Matthew Fontaine Maury, during his period of office as Superintendent of the Depot of Charts and Instruments in the Navy Department (1842–60), used techniques of statistical presentation in numerous maps prepared from data contained within the records in his care. One particularly effective distribution map from Maury was his *Whale Chart of the World* in which he used pictorial symbols to differentiate four types of whale. A rectangular grid is superimposed on the map and Maury's explanation states that 'Two Whales of the same species in a square denote that square to be much frequented by that species.' His most important maps, however, are his maps of the oceans showing winds and currents. These were compiled from information received from mariners throughout the world. In quite another field, Dr Edwin Leigh prepared an interesting map showing the percentage of illiterate adults by states in 1850 and 1860. Leigh used a technique which is now known as the *multiple unit* method in which one symbol or unit is assigned a certain value and a number of these units are then built up into a composite block symbol to represent the appropriate value at a certain point or within a given area.

The first official national census of the United States took place in 1790 but a long period was to elapse before any cartographical advantage was taken of the rich source of varied data provided by the decennial censuses. The first map to appear in an official report of the US Census was a small map showing drainage basins and though maps based on census data were prepared to show the distribution of slave population in the southern states, the census data was not fully exploited until Francis Amasa Walker published the *Statistical Atlas of the United States* in 1874 with maps showing physical and cultural distributions compiled from 1870 census data.

LAND-USE AND SOIL MAPS

Although quantitative developments tended to dominate the nineteenth century as far as thematic mapmaking was concerned, steps forward were also being made in qualitative mapping, particularly by earth scientists. In 1793 the Board of Agriculture and Internal Improvement was established, with Arthur Young as its secretary, and quickly organised a series of investigations into the agricultural state of different parts of Britain. The results of these enquiries were incorporated into a series of county reports, many of which contained hand-coloured maps. Young's *General View of the County of Norfolk*, for example, includes a map distinguishing loams, clay, peat, light soils and good sand, and the map in John Middleton's *Views of the Agriculture of Middlesex* (1798) features colours representing arable, pasture and nursery land.

One of the most ambitious maps published at the turn of the century was *Milne's Plan of the Cities of London & Westminster . . .* (1800) by the estate surveyor, Thomas Milne. The map was based on trigonometrical survey and printed at two inches to one mile. On it Milne recorded the results of a field-by-field land utilisation survey by means of a system of key letters and hand colouring in each parcel of land. Milne was of course following the normal practice of contemporary estate surveyors in showing land use but had developed his technique sufficiently to be able to differentiate seventeen types of land use as well as to make a distinction between common and enclosed land. His map was a forerunner of the modern land utilisation surveys of Britain made first by Professor L. Dudley Stamp and more recently by Alice Coleman.

GEOLOGICAL MAPS

Almost three centuries ago, in 1684, Martin Lister published *An Ingenious Proposal for a new sort of Maps of Countrys* in the Philosophical Transactions of the Royal Society. The kind of map he envisaged was a geological map but his suggestion was taken no further and he left the making of the maps 'to the industry of future times'. In 1725 J. Strachey published two sections which recorded his observations on rock strata in Somerset coalmines and in 1797 W. G. Maton issued his *Mineralogical Map of the Western Counties of England* which differentiated mineralogical variations in the rocks by means of line shadings.

It was at the close of the eighteenth century, however, that there occurred what has been described as the greatest advance in geology at any time in the history of the science. William Smith, born of yeoman stock in 1769, worked as a mineral surveyor investigating problems concerned with water supply, land drainage, the building of canals and the sinking of coal pits. During the normal course of his work he made numerous observations about rock strata and from them he arrived at the two conclusions on which his fame largely rests. First, he noted that the strata always succeed one another in a defined order and, secondly, that those fossils found in one particular stratum are peculiar to it – it followed that strata are identifiable from the fossils they contain. Smith incorporated these findings into maps; his geological map of England and Wales with parts of Scotland was published by John Cary in 1815 at five miles to one inch and was printed from fifteen engraved copper plates, the geological information being superimposed on a topographical base specially prepared by Cary, as Smith regarded the standard topographical map as too detailed for his purpose. Again we are brought back to the fundamental principle of good cartographic communication being dependent on the exclusion of distracting 'noise'. Smith, like Humboldt in his isothermal diagrams, was clearly conscious of the need to communicate his message strikingly and unhindered by irrelevant information. The outcrops of rock on Smith's map are separated by fine broken lines, with hand colouring applied to the outcrops in such a way that the tint appears darker at the base of the outcrop and shades off into a lighter tone over the remainder. Smith's map was the fruition of a quarter of a century's research and the various techniques he used to illustrate rock types on the basis of age and lithology formed the prototype from which modern geological mapping conventions have developed.

During the nineteenth century geological mapping was carried on vigorously and by the second half of the century most of the countries of Europe had detailed geological maps. These surveys were co-ordinated in the *Carte Géologique Internationale d'Europe* at a scale of one to one and a half million, a map which was one of the first important cartographic publications made with the co-operation of several nations.

THEMATIC MAPS IN THE MODERN WORLD

Thematic maps fulfil a vital need in modern society, particularly as a teaching aid at all levels of education and as a tool for spatial analysis by researchers in widely diverse fields. The vast number of statistical maps in atlases, text books and scientific journals provides a form of data bank from which a fund of information about the relationship between man and his environment can be drawn.

Much of the current output of published thematic maps comes from the cartographic units established in higher educational establishments since World War II and from government departments and local authorities.

Automated processes are increasingly being applied to the mapping of statistics for, as has already been mentioned, the computer offers distinct advantages in the rapid processing of data and its transformation into map form. Computer mapping will inevitably assume even greater importance as more data banks, in which information is stored on tape, are established. Providing that co-ordinates for outline maps of the relevant study areas are available on tape, and that co-ordinates of the data points can be produced, the possibilities of providing 'instant maps' from any kind of data are infinite.

11 OFFICIAL MAPMAKING TODAY

The immense output of every kind of map and atlas which pours forth from the presses of national mapmaking bodies makes it impossible to do more than draw the reader's attention to a few of the varied types of map and to selected examples of the establishments which produce them. Luckily the subject is well catalogued; the monumental catalogue of maps issued by Geo Center of Stuttgart is regularly updated and provides a useful general reference tool while the catalogues produced by individual establishments offer even greater detail of their publications, often including samples of maps at varying scales and index maps to each series. In addition to such catalogues there are standard reference works by G. Muriel Lock and Kenneth Winch which examine the entire field of modern cartographic production in some detail.

TOPOGRAPHIC MAPS

Apart from the ubiquitous weather map which confronts us daily and the constantly used automobile map, the most familiar map to the layman is likely to be the general-purpose topographic map, a map which, within the limits of scale, endeavours to provide as accurate a picture of the earth's surface as possible. It illustrates natural features – hills, rivers, forests, moor, heath, marsh – and those features added by man – railways, towns, roads, villages, canals, airports, industrial complexes and many more. A topographical map should be an indispensible adjunct to any journey and the true map lover will find it incomprehensible that anyone could embark on a visit to new territory without a map in his pocket or at least without consulting one prior to setting out. Furthermore, the topographical map is a vital working tool of geographers, botanists, archaeologists and workers in a number of other disciplines but particularly of earth scientists and social scientists. C. P. Snow, in *Strangers and Brothers*, wrote, 'I'm only just beginning to realise', said George, 'what a wonderful invention a map is. Geography would be incomprehensible without

maps. They've reduced a tremendous muddle of facts into something you can read at a glance.'

Up-to-date, accurate topographic maps are of great importance in war-time and it is no coincidence that the official survey and mapmaking bodies of several states, Italy for example, are military organisations.

It has been suggested that the modern period of cartography dates from a proposal made in 1891 by a German geographer, Dr Albrecht Penck, for an international map of the world at the one to one million scale. It was felt that topographic knowledge gained from centuries of exploration and discovery had reached such a level as to justify the production of a world series at this scale. Nevertheless the difficulties to be overcome in the production of an international series of this kind are acute; a suitable projection has to be found; conventional symbols developed which will cover a wide range of phenomena in diverse terrain and in countries at many different stages of development; a contour interval has to be chosen which will be equally effective in mountainous regions as in areas of lowland; a range of layer colours must be devised which will be as meaningful in temperate regions as in those of hot desert. The International Map of the World (IMW) was intended to be a general reference map which would also serve as a base for the mapping of other distributions. Initially its production was slow and World War I almost brought it to a standstill. Steady progress was made, however, between the wars, and national agencies were charged with the responsibility of preparing sheets of their own territories. Britain, for instance, produced 136 sheets of India and 132 of African territories, while the American Geographical Society made a major contribution with the fine *Map of Hispanic America* in over 100 sheets.

Today the IMW has become out-of-date; when the specifications were originally drawn up, the transport and communications system was very different to that of today. Aircraft and radio were virtually unknown and the automobile was only just beginning to appear. As far as general topographic information is concerned the problem is now one of deciding what to omit rather than what to include. Furthermore, the projection is no longer adequate for scientific research as the sheets do not fit and the general design is out of keeping with modern standards.

The rapidity of aeronautical development during World War II necessitated the rapid production of specially-designed navigational charts for use by airmen. The International Civil Aviation Organisation was established to this end and, as a beginning, numerous one-millionth scale sheets were placed at its disposal by the United States Air Force. From this beginning the current ICAO World Aeronautical Chart Series evolved. One essential difference between it and the IMW was that the latter was intended primarily as an indoor study map, while the former was a practical map for navigation purposes. A projection had to be found which was more suitable for navigation. That chosen was the Lambert Conformal Conic projection with two standard parallels, a projection commonly used for aeronautical charts due to its small scale error and its relatively straight azimuths over some hundred miles. The requirements of the aeronautical chart differ in other ways than that of the IMW; as it is designed to present the earth as seen from the air it is necessary to display towns according to the shape and extent of their built-up areas; international boundaries are of major importance to the aircraft pilot but minor internal boundaries are irrelevant; the ground communication network of roads, railways and canals is important to the airman only from its value in providing landmarks; names, except for those of major places, rivers and mountains are of little value. The airman does, however, need some additional information about airfields and their facilities, about radio beacons and air control data. When flying by night he requires such vital data to be provided in a colour which is clearly visible in the amber light of the aircraft cockpit. The ICAO series has been more useful than its predecessor, the IMW, but as aircraft attain faster and faster speeds further re-appraisals become necessary, and a few years ago the aeronautical navigational chart at a scale of 1:2,000,000 came into being.

The IMW and ICAO series were both confined to the mapping of land areas only, and in 1956 the idea of producing a map series which would represent the entire earth, including seas and oceans, at a scale of 1:2,500,000 was raised before the Economic and Social Council of UNO. At this venue the proposal was rejected but it was taken up by the combined geodetic and cartographic establishments of seven socialist states (Bulgaria, Czechoslovakia, Poland, Hungary, East Germany, Rumania and the Soviet Union). The interna-

tional co-operation involved in preparing the project was unique in cartographic history. It was a genuine co-operative venture which combined the fruits of joint discussion, collective compilation, collective appraisal of proofs and so on. This gave greater assurance of uniformity of style and a uniform quality of production which was lacking in the IMW, due to the number of countries involved and their varying standards of reproduction. The World Map 1:2,500,000 provides general information about the entire surface of the earth and serves as a framework for thematic mapmaking. Coastlines, hydrography, relief and other physical elements are shown and in the areas of economic and political geography are settlements, communications and boundaries. One problem encountered in the presentation of relief was that the available data from Anglo-Saxon areas was given in feet and, because the new map series was a metric one, it was necessary to interpolate contours in metres. Another problem, inevitable in such a project, concerned the transliteration of placenames. It was decided to name each sheet in English and in Russian, and this applied also to the legend and other ancillary information. On the maps themselves all names are given in Latin characters except for those of oceans and seas outside territorial waters, the names of which are given in Russian and English.

NATIONAL SURVEYS IN EUROPE

The remarkable Carte Géometrique de la France prepared by the Cassinis at a scale of 1:86,400 during the late eighteenth century served as model and inspiration for other European states, and the nineteenth century saw national surveys being undertaken by a number of state organisations. Britain's Ordnance Survey, originally the Trigonometrical Survey, was founded in 1791, its initial task being the scientific triangulation of the country, followed by the production of a national map series at the one-inch-to-one-mile scale. In its early stages, in which mapping was concentrated on the south-eastern counties, the survey fulfilled military requirements as there was considerable anxiety at this time about the possibility of invasion across the Channel during the Napoleonic Wars. The first one-inch map was published in 1801 and entitled *General Survey of England and Wales: an entirely new and accurate survey of the county of Kent, with part of the county of Essex.* The interesting point about this pioneering sheet was that it was privately published by William Faden and designed as a county map in the tradition of private mapmaking. It was not until 1805 that the sheet officially designated Number One of the Old Series or First Edition of the one-inch map was published. Thereafter the series was

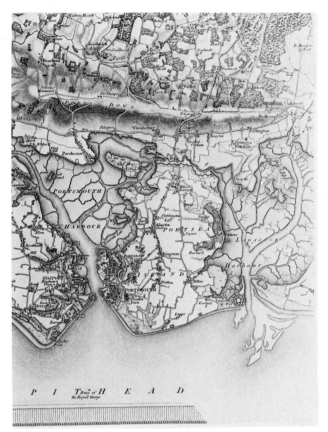

Fig 109 **Detail from Ordnance Survey First Series One Inch to One Mile, Sheet 11 (1810), with beautifully-engraved hachuring, form lines around the coast, woodland and varied topographical detail.**

was combined with orange contour lines to produce maps of considerable aesthetic appeal.

To many people an OS map has always meant a one-inch map or, today, a 1:50,000 sheet. The Survey, however, publishes several other series, ranging in scale from 1:1,000,000 to the ultra-large-scale urban plans published during the nineteenth century at 1:500 (10.56 feet to one mile), and including the 1:2500 plans which originated from surveys made during the period 1853–98. Although the original 1:2500 sheets on county sheet lines have been replaced by plans on National Grid sheet lines, the present sheets at this scale are still the most detailed maps available for most of Britain. Only in the larger urban concentrations, where an even larger scale (1:1250) is necessary, and in areas of mountain and moorland where the smaller scales of 1:10,000 or 1:10,560 are adequate, has the 1:2500 been replaced as the basic series. The scale of 1:2500 was in accordance with the recommendation of an international statistical conference held in Brussels in 1853 concerning the most suitable scale for a cadastral survey, and prior to metrication it had an additional asset in Britain in that one square inch on the plan represented almost exactly one acre on the ground. The aim of such a large-scale series is to depict permanent features of the landscape as faithfully as possible, and the sheets are true plans in that no deliberate distortion of scale in order to insert names, or for any other reason, is involved. The 1:2500 sheets are of great value in conveyancing and land registration because of the system by which each parcel of land is numbered on the map and its area indicated. The series is also an essential tool of local authorities because of its large scale which makes it admirably suited to the addition of sewerage schemes, proposed roadworks and housing schemes and because of its very detailed depiction of local authority boundaries.

Apart from the one-inch series the six-inch (1:10,560) is the oldest OS map series, this scale being adopted in 1840 for areas of northern England and Scotland which were still unsurveyed. The six-inch series has two major distinctions; it (and its successor the 1:10,000) is the largest scale to cover the whole of Great Britain and the largest scale on which contours have been shown. Like the larger 1:2500, the six-inch series is particularly useful in local government, planning and civil engineering, while the early-nineteenth-century sheets are indispensible to the industrial archaeologist in their detailed indication of industrial sites which have since been demolished.

The 1:25,000 series grew out of a recommendation of the Davidson Committee in 1938 that 'a new medium scale of 1:25,000 should be tried out experimentally in

extended to encompass the whole of England and Wales, a task finally accomplished in 1873 with the publication of the last of the 110 sheets.

The story of the one-inch series, now closed, with its replacement by the metric 1:50,000 series, is of slow progress involving important developments in surveying instruments and methods, as well as in the techniques of map reproduction. It provides an engrossing insight into the evolution of the distinctive cartographic style which makes an Ordnance Survey sheet immediately recognisable. The Old Series sheets were printed in black and white only from engraved copper plates, their most distinctive feature being the hachuring used to depict hills and which gave mountainous areas such as Snowdonia a heavy, black appearance. By the last decades of the nineteenth century, map reproduction techniques had become much more sophisticated with the introduction of lithographic printing, photographic methods and colour printing. Each series of the one-inch map had its own individuality but few will deny the especial charm of the Fifth (Relief) Edition in which delicate grey hachuring

certain selected areas, and if successful, should be extended to cover the whole country'. The recommendation arose from the widely expressed opinion that there was too considerable a gap between the existing one-inch and six-inch scales and that, for educational purposes in particular, the former was too small to give the required detail while the latter was too large to give a general picture of the country and, furthermore, was costly and presented storage problems if coverage of any sizeable area was to be maintained. It was also felt that the map would be welcomed by the general public as a walking map and indeed the 1:25,000 is ideally suited for practical use outdoors or as a study map. Its most distinguishing feature from the larger scales is the use of colour.

The current 1:50,000 series came into being as a replacement for the Seventh Series one-inch maps, its main advantages being first that this scale was already widely used in other countries, particularly in Europe, and second, that the scale could readily be expressed as a simple centimetres to kilometres relationship. The 1:50,000 series was introduced in two stages, a First Series, on new sheet lines, which was basically a photographic enlargement of the Seventh Series one-inch map and a Second Series which would be a completely redesigned map. The larger scale (approximately one and a quarter inches to one mile) has undoubtedly given greater clarity and there are a number of changes in the use of colour and in symbol design. Contours, for example, are now shown in orange, rather than the brown of the Seventh Series, while woodland is now shown in green without the tree symbols which formerly distinguished deciduous and coniferous. Another rather controversial change, but one which is successful in enhancing the legibility of superimposed detail, is the substitution of an orange tint for built-up areas rather than the black tint of the one-inch, but the most surprising departure from convention is the use of a strong blue colour for motorways. The choice of blue was made to accord with the blue signs used on motorways, a convention which had already been used successfully in Europe. In order to make the 1:50,000 a truly modern map it has been necessary to provide new symbols which relate to the increasing number of leisure activities. Revised administrative boundary information was essential as a result of the Local Government Reorganization Act, 1972. Changes have also been made in the typography in order to update the appearance of the Second Series and the well-tried Gill Sans type face, first introduced by the OS in the 1920s, has been replaced by what is perhaps the most popular face in use today, Univers. All in all, the 1:50,000 is a satisfactory successor to the Seventh Series, with an improvement in legibility and an increased amount of detail which reflects the new emphases of contemporary life on motoring, on recreation and on the increasing number of foreign visitors to Britain who are catered for by a legend in French and German as well as English.

The Ordnance Survey caters for the motorist with a 1:253,440 (quarter-inch to one mile) series and its metric successor, the 1:250,000, though these maps meet with stiff competition from commercially-produced motoring maps, particularly the modestly-priced oil company maps which the motorist can conveniently purchase when he calls at a filling station.

As well as fulfilling a statutory obligation to include administrative boundaries on its main map series, the Ordnance Survey publishes various special Administrative Maps. The whole of Great Britain is covered at a scale of 1:625,000 in two sheets which for England and Wales show administrative counties, county boroughs, London boroughs, municipal boroughs, urban districts and rural districts; for Scotland, counties of cities, burghs and district councils; for the Isle of Man, borough and town. These sheets of course show the situation before the Local Government Reorganization Act of 1972, an act which illustrates some of the problems of a national mapping organisation, for at one stroke it rendered existing maps out-of-date. Administrative maps on county sheet lines at quarter-inch and half-inch scales have been issued since 1879. These series were available in two styles; the first showed in red, on a grey base, county and local authority boundaries and in green, parliamentary counties, county divisions, parliamentary boroughs, etc.; the second style showed only local government boundaries in red on a grey base. The first sheet of a 1:100,000 administrative series was published in 1965 and shows the following information: administrative counties, county boroughs, municipal boroughs, urban districts, rural districts, rural boroughs, civil parishes, county and borough constituencies. A second set of 1:100,000 maps depicts petty sessions areas.

Before leaving the Ordnance Survey, mention must be made of the series of thematic maps at 1:625,000. These were prepared in collaboration with government departments which contributed in large measure to research and compilation. An extensive range of themes was covered and the series forms Britain's nearest approach to an official national atlas. At the present time the Ordnance Survey is responsible for the official surveying and mapping of Great Britain. One of the major tasks is the updating of nineteenth and early twentieth century 1:2500 surveys, another is that of making entirely new surveys of main towns at 1:1250. Moorland and mountain areas are to be mapped at 1:10,000 with a

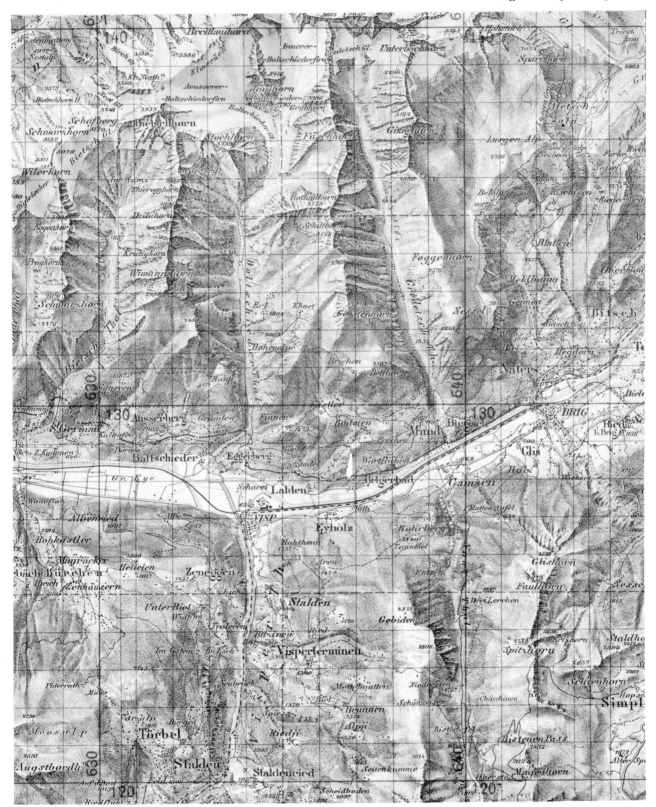

Fig 110 Detail from *Carte Topographique de la Suisse, Carte Dufour*, 1:100,000, showing part of the Rhône valley with a vivid three-dimensional impression of the side valleys obtained by superb drawing of hachures.
Reproduced with the permission of the Topographical Survey of Switzerland from 31 January 1977

target date for completion of 1980. Revision in these days of rapid change is a formidable undertaking and new maps are kept as up-to-date as possible by a system of continuous revision which ensures that any changes on the ground are quickly surveyed.

Visitors to European countries often rely for maps on the commercially produced series from well-known names such as Hallwag, Michelin, Firestone, Kummerley and Frey or Freytag and Berndt. There is little general awareness among the general public outside Europe of the wealth of maps issued by official cartographic establishments, largely because their publications are not so readily available via retail outlets. Most of the European nations have main map series at 1:25,000, 1:50,000, 1:100,000 and so on but there is no absolute standard of style, content or quality of production. Individual nations have maps in recognisably different styles just as they did in the heyday of engraved maps and a comparison of maps at the same scales will reveal something of the varying approaches adopted to similar tasks.

Clearly a quite different system of conventional symbols will be needed in the maps of countries as diverse in terrain as Switzerland and the Netherlands. A glance at the legend of the Netherlands 1:50,000 series will reveal symbols for windmills; watermills; motorised wind pumps; dikes of varying heights; numerous items associated with canals; sluices; and tramways along roads, as well as more universal symbols such as those for churches, railways and boundaries. Contours are at 5m and 10m intervals, for in a flat country close contouring is essential to display the relief forms. In complete contrast is the mountainous terrain of Switzerland which lends itself to the making of visually attractive maps. The contours of the Swiss *Carte Nationale* are at 20m intervals, a choice which provides a vivid picture of the mountainous landscape, particularly when combined with a skilled use of shadow. The Swiss National Survey, *Eidgenössische Landestopographie*, founded in 1838, maintains a long tradition of beautiful mapmaking in which great care is exercised in the choice of type faces, symbols and colour schemes to ensure maximum clarity and effectiveness of communication. It is, of course, in the depiction of mountain and glacier landscapes that the Swiss excel. The delicate brown hachuring of the oldest Swiss series, the *Carte Dufour* at 1:100,000, is a *tour de force* which, without the aid of shadow, evokes a striking impression of rugged terrain. Swiss rock drawing is a fine art and is seen to perfection in the *Carte Nationale de la Suisse*, also at 1:100,000. The graphic impression of relief in this series is achieved by brown contours allied to grey shadow, and a restrained use of colour – roads, for example, being left without colour infilling. On the 1:50,000 maps too, the absence of the conventional red roads of the corresponding Netherlands series adds surprisingly to the clarity without any loss of visual appeal. Swiss maps display other significant differences from those of the Netherlands. The transport system includes single and double track railways, cable railways, funiculars, and road classifications ranging from mule tracks and tracks over glaciers to first-class roads. Bobsleigh runs are possibly unique to Swiss maps.

The landscape of the Netherlands hardly seems likely to inspire the making of such beautiful maps but the country has an unrivalled cartographic tradition and is one of the modern world's leading mapmaking nations. The Netherlands have outstanding achievements in the academic teaching of cartography, in the design of cartographical instruments and equipment, and in the fine map series of the national agency, *Topografische Dienst* of Delft. Dutch maps, produced by sophisticated technology, convey an extensive range of information without overcrowding, one of the hallmarks of successful cartography. The most immediate impression given by most Dutch maps is of a dense network of roads, an impression heightened by the strong red of the main roads, and an overall pattern of blue drainage ditches, canals and rivers. The line of dunes backing the coastline is nicely shown in brown hachuring and in older towns like Haarlem the number of churches is a notable feature.

The cartographic organisation of the Federal Republic of Germany differs from the usual pattern of centralised survey. In Germany, cartography is largely the responsibility of land-survey administrations in each of the *Länder*. These bodies look after the large-scale mapping of their own territory as well as some small-scale work. Such decentralisation inevitably leads to problems in achieving standardisation and, in order to handle discrepancies between mapping organisations, a working group composed of members from federal bodies and from each *Land* was inaugurated. This group also dealt with the general problems of preparing small-scale maps, for the Land Survey bodies were too occupied with urgent tasks of post-war large-scale survey needed for re-development to be able to cope with additional small-scale work. An agreement made in 1952 decided that the Institute for Applied Geodesy should prepare such maps, and series have been published at 1:200,000, 1:500,000 and 1:1,000,000. Two of the major German achievements in post-war cartography have been the preparation of the multi-colour *Topographische Karte* at 1:50,000 and the revision of the 1:25,000 series. A shaded relief edition of the former has been published for the southern part of the country in order to meet the needs of ramblers and tourists. A comparison of sheets from two of the Länder pinpoints the difficulty of achieving uniformity of style. The *Topographische Karten* at 1:25,000 and 1:50,000 from Schleswig-Holstein and Saarlandes are markedly different in appearance. The greens used for woodland and the browns of the contours are at variance. The Schleswig-Holstein roads have a red infill on the 1:50,000 sheets whereas those of the Saarlandes do not. The legends to these series, however, include an impressive amount of cultural and physiographic information and, despite a tendency to use rather old-fashioned typography, the maps are clear and legible.

The *Istituto Geografico Militare*, established in Florence in 1872, keeps up the traditionally high standards of Italian mapmaking. The incomplete 1:50,000 map series is an attractive one with close contouring, effective use of shadow and skilled rock drawing. A wealth of detail is included. The communications section of the legend includes symbols for streetcar lines, trolley bus routes, funiculars, chair lifts, ski tows and cable railways as well as more conventional road and railway information. Oil and gas pipelines are differentiated according to their situation – underground, surface or elevated. Walls, retaining walls and dry masonry walls are given separate symbols but it is in the indication of land use that the 1:50,000 series is especially rich, with eighteen different classifications including vineyards, orchards, citrus groves, firs, scrub and afforestation. The 1:25,000 Italian series appears in three alternative versions: in black; in three colours; in five colours. These provide some indication of the way in which the series has developed. The black and white sheets of the original series were superceded by three-colour maps in 1946 and these in turn by five-colour sheets in 1958. A comparison of sheets in each of the three styles reveals the improvement in clarity when five colours are used. The distribution of woodland, for example, in the earlier series is difficult to appreciate but is seen at once when picked out in green on the five-colour sheets. The *Carta d'Italia* at 1:100,000 is available in three styles: with relief by contouring only; with brown shadow added; in an administrative edition. The relief depiction of this series differs strongly from that of the 1:250,000 regional maps. The 1:100,000 has an overall brown appearance while the 1:250,000, with both contours and shadow in blue, gives a predominantly blue impression. In neither case does the colouring relate closely to the terrain under consideration.

What of the Scandinavian countries which lack the long tradition of fine mapmaking enjoyed by Germany, Holland and Italy and which have such a variety of landscape from the gentle hills of Denmark to the glaciers of Norway and the forests of Sweden? Denmark's first mapping agency, *Videnskabernes Sels-kab* (Royal Danish Society of Science and Letters), began to operate in 1757, but in 1808 the Danish General Staff began to build up a maintenance service for military purposes and in 1842 a Topographic Branch was inaugurated to assume responsibility for topographic mapping in Denmark and her overseas lands. The establishment in 1816 of *Den Danske Gradmaaling* (Royal Danish Arc Survey) was a parallel development. In 1928 the two bodies united to form the agency now responsible for all Danish official mapmaking, the *Geodaetisk Institut*. Current series range from the 1:20,000 covering the country in 835 sheets, to regional maps at 1:300,000 and 1:500,000. The 1:20,000 is printed in only three colours and, in principle, shows all topographical detail which can be drawn true to scale, except for items like roads, railways and historical features, which are shown by conventional symbols. As Denmark is relatively flat, contours are inserted at every 2 or 2.5 metres, with spot heights added to make interpretation easier. The 1:25,000 series shows the same detail as the 1:20,000 but is printed in six colours. The 1:50,000 is derived from the 1:25,000, has the same six colours, and provides essentially the same information except that the exigencies of the smaller scale mean that only the largest buildings are true to scale, others being symbolised. On the older sheets, suburbs and villages are represented by an arbitrary number of house symbols to convey the character of the area. On newer sheets a light grey tint is used. Denmark caters for the motorist with a 1:200,000 series in seven colours which is derived from the standard larger-scale series, except that relief depiction is limited to strategically-located spot heights. Cadastral maps at 1:800 and 1:400 are issued by the Land Register Survey.

Sweden has three main cartographic establishments: *Generalstabens Litografiska Anstalt* (GLA), Esselte Map Service and *Rikets Allmänna Kartverk*. Sweden has earned an excellent reputation for a forward-looking approach to topographic mapmaking and has also gained renown in the field of statistical mapmaking, the economic and population maps of Professor W. William-Olsson being outstanding examples of their kind.

Norway's official agency, *Norges Geografiske Oppmåling*, publishes topographical series at 1:25,000, 1:50,000 and 1:100,000 as well as handsome tourist sheets such as that of the *Hardangervidda* at 1:200,000, a sheet which displays Norwegian official mapmaking at its best. The close contouring in brown, unaided by shadow, shows the high mountains rising steeply from the edges of the fjords – what a vivid picture is presented, for example, of the narrow Eidfjord and the valleys leading from it. The few roads stand out well in their

Fig 111 **Part of a large sheet map of Hong Kong at a scale of 1:50,000. In order to meet the needs of Chinese as well as English-speaking users, the names are provided in both languages.**
Reproduced with the permission of the Crown Lands & Survey Office, Hong Kong Government

deep red and so does the drainage pattern in blue. Items of especial Norwegian interest are seen in the legend. The all-important ferries, for example, are represented by a rather ineffective symbol consisting of a rectangle flanked by two short lines on each side; farms are symbolised by an open circle, *saeters* by a black dot. The importance of outdoor recreation is evident in the inclusion of hotels, fishing and shooting huts, tourist centres, ski-tows and a network of well-marked footpaths and tracks. As in some national series already mentioned, the overall colouring of the Norwegian 1:20,000 bears little relation to the terrain represented, except perhaps for the ice-caps which are left white as a contrast to the buff tint of the remainder of the map. The three-dimensional shaping of the landscape, however, is well brought out.

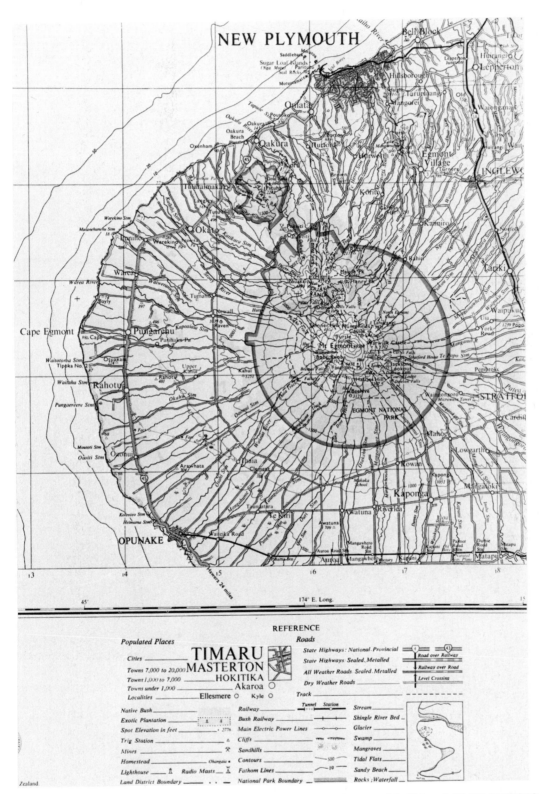

Fig 112 Mt Egmont is strikingly portrayed in this detail from the New Zealand Topographical Map, 1:25,000, NZMS 18 Taranaki, Sheet 7, 1st edition, 1970, published by the Department of Lands and Survey, New Zealand. The legend gives a good illustration of the way in which varied sizes and styles of type are used to convey the status of towns and cities.
Reproduced by permission of the Department of Lands and Survey

FALKLAND ISLANDS DEPENDENCIES
SOUTH SHETLAND ISLANDS

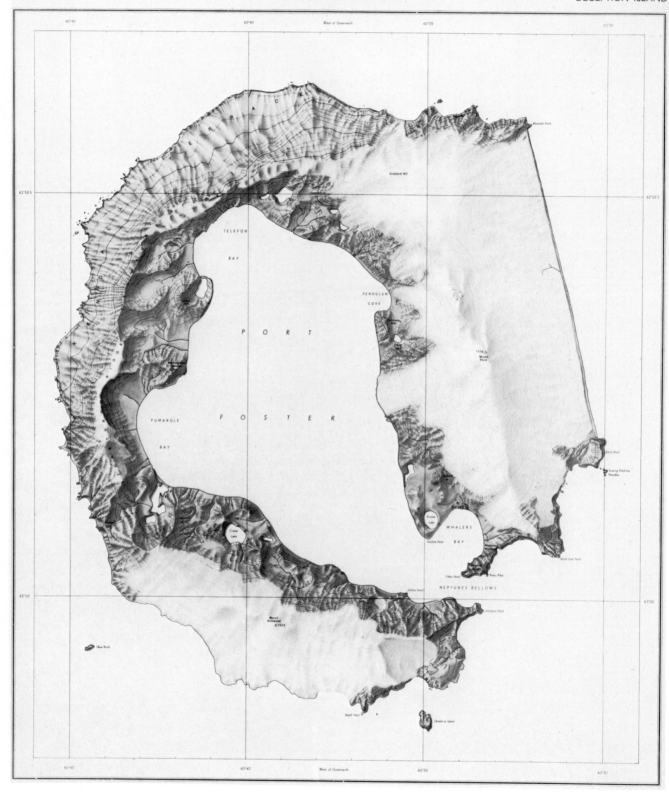

Fig 113 **The Directorate of Overseas Surveys (DOS) in its mapwork for former British colonial territories and developing countries produces some of the most exciting modern cartography. The illustration, at a much reduced scale, conveys a splendid impression of the topography of Deception Island as well as being a pleasing piece of graphic art.**
© Crown Copyright
Reproduced from Directorate of Overseas Survey's Map DOS 310, Deception Island, Edition 1, by permission of the Controller of HM Stationery Office.

Limitations of space preclude further examination of European mapmaking but this does not belittle the cartography of countries such as Spain, France, Portugal or Austria which produce excellent map series. The achievements of countries within the Soviet bloc and of the USSR are difficult to assess owing to the problems of procuring sufficient examples and of ascertaining full details of scales and map series. The evidence available, largely in the form of sheet maps of tourist areas, plans of major cities, and the 1:2,500,000 world series, indicates that the nations of Eastern Europe occupy a place well to the forefront of contemporary world cartography.

Before turning to American cartography, the achievements of former British colonial territories must be recognised. India has imposing traditions dating back to 1767 when Major James Rennell was appointed Surveyor General of Bengal. A country of such varied terrain, climate and land use presents a challenge to the mapmaker, particularly in devising a cartographic symbolisation which applies equally well to the Himalayan regions as it does to the Ganges plain or the arid regions of Rajputana. Items peculiar to India which are seen on her official map series include some connected with local government administration such as dak bungalows and rest houses, while the variety of gauges in the Indian railways system meant that the mapmaker has resorted to writing the gauge alongside the line itself. Like India, Egypt is well covered at varying scales and has built on the British tradition.

The organisation of cartography in Australia differs from that of other nations, in that Federal Survey is in the hands of the Director of National Mapping, Department of National Development, but separate states have their own mapping authorities which prepare large-scale administrative and cadastral maps, together with general-purpose maps and tourist sheets. The Bureau of Mineral Resources, Geology and Geophysics is responsible for geophysical and geological mapping in Australian territories, or within in an individual state, if invited by the state authorities.

The official mapping organisation in New Zealand is the Department of Lands and Survey which issues a topographical series at one inch to one mile, as well as smaller-scale series, special-purpose maps for leisure and recreation, and a number of cadastral map series. The 1:63,360 series is particularly attractive and its effective relief portrayal is achieved, not by close contouring (the interval is only 100 ft), but by skilled use of shadow allied to clear contour lines. Bush and scrub are shown in green and National State Highways in red, and there is one particularly unusual feature – the inclusion of a symbol for burnt or fallen bush. There is a special emphasis on bridges, seven types – two-lane, one-lane, concrete, wooden, steel, suspension and footbridges – being indicated. The one-inch series, like that of the British Ordnance Survey, has outlived its usefulness and is to be replaced by a metric 1:50,000 with a closer contour interval of 20 metres. An unusual feature of the New Zealand survey is the number of cadastral series, including one at 1:50,000 showing land section boundaries, section or lot numbers, Maori block boundaries and names, deposited plan numbers and other administrative information. Cadastral maps for towns show similar information at larger, scales, varying from 1:4752 to 1:10,000.

Canadian maps are unquestionably among the world's finest, admirably designed, distinguished in their typography and beautifully colour-printed. The making of maps is controlled by the Department of Energy, Mines and Resources and the National Topographic System (NTS) includes major series at 1:25,000, 1:50,000 and 1:250,000. In 1967 the whole nation was mapped for the first time at 1:250,000. Most developed regions are now covered at 1:50,000 and cover at 1:25,000 is provided for densely-populated areas. Special problems were encountered in producing the national 1:50,000 series and in 1972 a study was made of the use of this series in the northern part of the country. Here sales were largely restricted to geologists, prospectors, ecologists, petroleum engineers and transportation engineers. All these users were regarded as experienced map users who could cope with a less explicit map than the standard six-colour sheet. Furthermore, it was found that some users preferred a map which was not almost entirely covered by a green vegetation tint, for much of this area is forest or muskeg and the green tint pervades the map to the detriment of other information. On the other hand, a monochrome map, if not thoughtfully designed, contains many sources of confusion such as the similarities between streams and contours. In the event it was decided to produce single-colour maps in order to speed up production for an area north of the 'Wilderness Line'. In such an area of streams, lakes and forests the mapmakers paid particular attention to drainage and related features, and the legend includes items varying

Fig 114 **Peyto Glacier was one of the glaciers surveyed and mapped for mass energy and water balance studies as part of Canada's contribution to the International Hydrological Decade (1965–74). The Peyto Glacier map, the first of its kind in Canada, presents information to interest earth scientists and rock climbers alike. It is valuable as a teaching aid and at the same time provides a travel guide for the tourist. Swiss expertise was called upon to assist the Canadian personnel in hachured bedrock portrayal. The cartography is by the Inland Waters Directorate of the Department of the Environment and the map is published by the Canada Map Office, Department of Energy, Mines and Resources, Ottawa, 1975.**
Reproduced by permission of the Minister of Supply and Services, Canada.

from tundra, rapids, marsh and inundated land to intermittent lakes and dry river beds. The communication network presented much less of a problem for there are hardly any railways in northern Canada and the road network is similarly sparse. Seaplane bases had to be symbolised, however, and a wealth of landmarks was necessary in this type of terrain – fire and radio towers, transmission lines, mines, oil wells, houses, barns and schools all being located. Like the new Ordnance Survey 1:50,000 series the Canadian maps do not distinguish coniferous and deciduous forests. The specification for forest is that all areas are shown where over 30 per cent of the ground is covered with trees at least six feet high. Another special feature was the indication of seismic lines, cut by bulldozers through forests during geophysical exploration. These were indicated partly for their value as landmarks and partly because of their potential as rough trails.

The topographical series of the United States are produced by the US Geological Survey, a body formed in 1879 to consolidate four earlier organisations which had been concerned in topographical and geological mapping. The importance of mineral deposits in the west led to geologists demanding an accurate base map on which

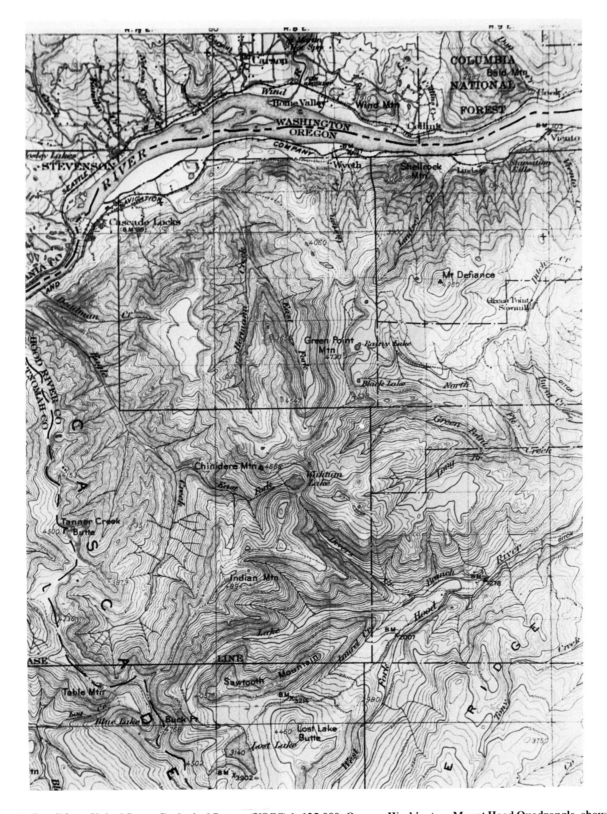

Fig 115   Detail from United States Geological Survey (USGS) 1: 125,000, Oregon-Washington, Mount Hood Quadrangle, showing
how well close contouring can convey the dramatic nature of terrain.
Courtesy: United States Geological Survey

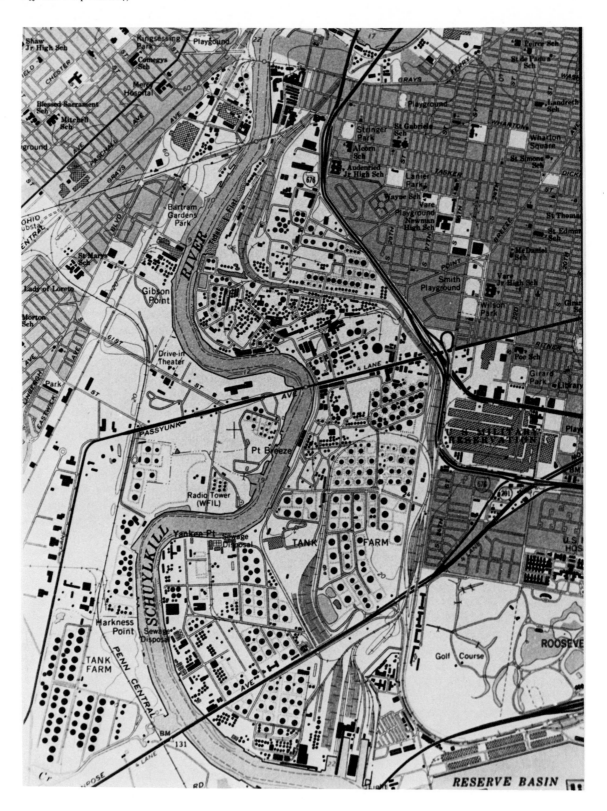

Fig 116 **Detail from United States Geological Survey, 1:24,000, Philadelphia Quadrangle, 1967, Photorevised 1973. A detailed but admirably clear portrayal of urban and industrial landscape at a medium scale.**
Courtesy: **United States Geological Survey**

they could superimpose their findings. A Topographic Branch of the Geological Survey was formed for this purpose. The US Geological Survey today has four operating divisions: topographic, geological, water resources, and conservation; and one of its major functions is the preparation and maintenance of topographic map series. The basic triangulation and levelling which form the basis of any map series have been, and remain, the responsibility of the US Coast and Geodetic Survey, originally called the United States Coast Survey. Initially this body's sole function was to delineate the coastline and produce hydrographic charts, but it was soon apparent that the coasts of the Atlantic and Pacific Oceans must be joined by trans-continental triangulation in order to secure their proper co-ordination. Work on this took place from 1871 to 1896. The interior triangulation was then found to be so valuable for other purposes than co-ordinating the coastal charts that the Coast Survey was made responsible for geodetic surveys throughout the nation and in 1873 its name was changed to Coast and Geodetic Survey.

The overall survey plan divides the whole country into quadrangles limited by parallels of latitude and meridians of longitude. These quadrangles are mapped at different scales, the scale of a particular map being that which is best suited to general use in the development plans of the nation. In areas where problems of outstanding national importance occur, surveys are of sufficient accuracy for maps at 1:31,680 with contours at one, five or ten feet intervals; where problems are of average national importance, the survey is accurate enough for maps at 1:62,500 with contours every ten or twenty-five feet; where the problems are of only minor national significance, survey accuracy suffices for 1:125,000 maps with contours at twenty-five to one hundred feet intervals. The colour symbolisation of Geological Survey sheets is planned to a well-defined pattern. Water features are in blue, man-made objects in black, wooded areas green, and symbols relating to the shape and elevation of the terrain in brown. An interesting distinction is made between buildings used to house human activities, and those designed for the protection of machinery, animals or materials. Special regional requirements call for the inclusion of specific individual items so that we find wells and springs shown in the arid western states and vineyards or mine dumps in appropriate areas. The wealth of cultural information, which reflects the generally advanced state of development, does not preclude clarity of presentation.

The contemporary emphasis on cultural detail contrasts with the early days of the Geological Survey when the greatest care and expertise was lavished on the depiction of the physical landscape. Even in the most recent maps, however, relief is finely executed and many of the sheets are highly spectacular merely because of the exciting terrain they display. Indeed it is true to say that no other nation, with the exception of Soviet Russia, has such varied problems to face in its production of topographic maps due to the complexities of the cartographic language which are involved when mapping at a continental level.

12 MODERN COMMERCIAL AND PRIVATE CARTOGRAPHY

Although original survey at national level has, since the nineteenth century, become almost a prerogative of governmental mapmaking organisations, the private sector of cartography is responsible for an immensely varied output of maps, normally in the form of compiled maps constructed on an official survey base. A copyright fee or royalty is normally charged for official data used in this way, though some nations, the United States being one example, deem maps to be in the public domain and permit free usage. The bulk of the commercial output consists of general maps, various kinds of atlases, route maps and educational material such as wall maps and globes.

COMMERCIAL MAPMAKING IN BRITAIN

Numerous commercial establishments, both large and small, are engaged in the private sector of cartography in Britain. Among the best known is the Edinburgh family firm of John Bartholomew & Son Ltd, an old-established firm with links dating back to 1797 when George Bartholomew began work as a map engraver in Edinburgh. Perhaps the map series most immediately associated with Bartholomew is the half-inch-to-one-mile (1:126,720) topographical series of Great Britain, a series which became widely popular as a general-purpose map, because of the clarity of its design and the effectiveness of its layer colouring, a method which Bartholomew introduced for the first time into a topographical series. The half-inch series is being replaced by new maps at 1:100,000 to bring it into the metric line and the additional clarity resulting from the larger scale should ensure its continuing popularity, particularly in leisure pursuits, for the series has long been a favourite of walkers, cyclists and motorists. No mapmaking enterprise these days can afford to ignore the motorist, and Bartholomews provide an excellent GT Tourmaster series of motoring maps, Ireland Travel Maps and several road atlases. Like many other commercial publishers, Bartholomews produce educational atlases

at varying levels ranging from works for younger children to the well-known *Times Atlas of the World* in five volumes. The Bartholomew World Travel Series offers useful coverage of the continents, regions and countries of the world at varying scales and at the lighter end of the firm's output is a series of pictorial maps on themes as diverse as whisky, football, stamps and the Spanish Armada.

George Philip & Son Ltd have always been biassed towards the educational field and have played their part in geographical teaching at all levels. Many English readers will remember being confronted in the classroom by a Philip's wall map or globe or using a Philip's School Atlas. The Clarendon Press of Oxford has also had a long association with the field of education and the joint educational publishing activities of Collins and Longman have produced numerous atlases of the United Kingdom and overseas territories.

The British firm of Geographia Ltd is especially associated with town plans and with maps designed to convey useful information to the businessman. Other undertakings such as Hunting Surveys Ltd and Fairey Surveys undertake a wide range of surveying and mapping projects on an international scale. Fairey Surveys, for instance, have been engaged on a Bahamas survey which involved some interesting operations. The land surveyors travelled extensively by boat and made use of infra-red photography to determine water-lines and to provide a datum for the contouring system. In some contrast to this rather exotic work, the firm has been involved in road construction mapping programmes in which detailed plans have been prepared at scales as large as 1:500. Numerous small cartographic firms have appeared in the post-war years. David L. Fryer & Co. are a good example of the smaller establishment which offers a range of work from the preparation of medium- and small-scale maps to atlases, road maps and town plans.

Another post-war development was the setting up of cartographic units in higher educational establishments,

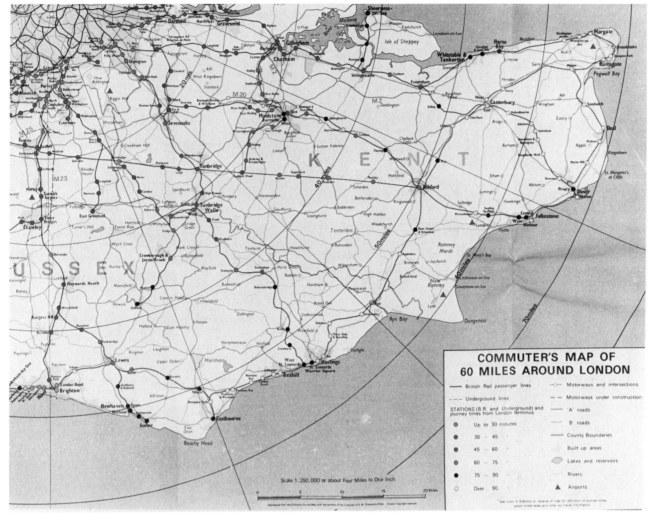

Fig 117 **Part of the** *Commuter's Map of 60 Miles around London* **published by John Swan & Co., London, at 1:250,000. The map shows British rail routes and underground lines with stations coloured to show the length of journey time into Central London. Motorways and 'A' and 'B' roads are shown and circles indicate the distance from the centre of the city. The reverse of the map has a table of railway stations with their appropriate London terminus, journey times, and cost of monthly season ticket. The illustration is from the original publication by Swan but the map is now published by Geographia Ltd.**

usually housed in Departments of Geography. In this sector, the emphasis is on thematic maps prepared as illustrations for textbooks and journals or as visual aids for teaching purposes. Despite the small number of personnel involved, such units play a significant part in the diffusion of geographical knowledge. Much of the work is reproduced at a very small scale in black and white. This means that the cartographer has to devise means of graphical presentation which will communicate an often complex message quickly and clearly. Much use is made of dry-transfer methods and photo-composition for lettering, while hand-shaded methods are combined with the use of pre-printed tint sheets for the in-filling of areas. Above all, the mapmaker must have a strong feeling for layout and design in order to convey his theme to the greatest effect. Computer-assisted mapmaking plays an increasing role in the educational sector, for the computer offers distinct advantages over manual methods in the processing and transformation of statistical data into map form. Numerous thematic atlases have already been produced with the aid of the computer and the quality of visual presentation is gradually being improved.

THE EUROPEAN SCENE

Commercial mapmaking flourishes in Europe with over forty private firms existing in Germany alone, an astonishing figure which indicates the German public's insatiable appetite for maps. The annual output of road maps

Fig 118 **Detail from a plan of Milton Keynes at a scale of 1:13,333 prepared by David L. Fryer & Co. of Henley-on-Thames. The plan is clear and well-balanced with an easy legibility achieved by the judicious use of a modern sans-serif type combined with more limited use of seriffed type for particular features. Tints of varying intensity have been used to show residential areas, industrial areas and buildings.**
Courtesy: **David L. Fryer** © **Milton Keynes Development Corporation**

in Germany is considerably over ten million and it is clear that the average German is more familiar with a motoring map than with any other form of cartographic output. It is the motoring map therefore which moulds his views of the cartographer's craft and from the industry's point of view it is important that road maps should be well designed as well as functional. One of the major commercial publishers in Germany is the Mair's Geographical Publishing House founded in 1950. Mair's major series is the *Deutsche Generalkarte* at 1:200,000 which covers the country in uniformly-sized sheets based

on a new survey and which is kept up-to-date with special emphasis on highway conditions. Mair also publish a number of road atlases of European countries in which the main maps are at a uniform scale, and maps of particularly important areas are presented at twice the scale to give even greater clarity. In order that the user can quickly identify routes of scenic or cultural interest, special attention is paid to the selection of conspicuous symbols to represent scenic and cultural features. Other establishments in Germany include JRO Verlag who publish a wide range of items – road maps, topographical maps, atlases, globes and panoramic maps of a pictorial character, which provide the viewer with an immediate three-dimensional portrait of the landscape; Westermann Verlag specialise in wall maps and atlases; Falk Verlag of Hamburg concentrate on town plans, road atlases and route maps; Geo Centre of Stuttgart have an extensive cartographic output, but one of their most useful products is their comprehensive catalogue of maps from agencies throughout the world. Prior to

World War II, the Justus Perthes Geographische Anstalt in Gotha was pre-eminent in Germany, but all firms were nationalised in East Germany after the war. The Gotha Institute became VEB Hermann Haack, an establishment which has maintained the Gotha traditions, particularly in the educational field, the Haack wall maps being instantly recognisable for their distinctive colour schemes and their concentration on clarity for classroom use.

Tourists in Europe can hardly fail to have met the Michelin series of maps and guides. André Michelin wrote his first guidebook expressly for motorists in 1900, following it in 1913 with the first Michelin road map. One of the best features of the Michelin guides (Green Guides devoted to tourist information and Red Guides concerned with hotels) is the quality of the maps and town plans. These are models of good cartographic illustration and are generally printed in two colours, brown or red for the map detail, with lettering in black. There is a natural concentration on highways but it is conceded that not everyone travels by car, and information about railways, tramways, trolleybus routes and so on is provided for the non-motorist. The main Michelin sheet map series are also aimed primarily at the motorist but provide information for others. Roads are shown clearly and very prominently on Michelin maps but the remainder of the map can seem rather cluttered, due to the abundance of names included. No attempt is made at relief depiction and the motorist studying his Michelin 1:200,000 of the Dieppe area is faced with a seemingly flat countryside instead of the pleasantly rolling reality.

Private cartography co-exists harmoniously alongside official survey agencies in most Western European countries. While it is impossible to mention more than a very few of the European commercial establishments, the visitor to Europe may well come across the attractive picture maps of Belser Verlag in Stuttgart, the perspective plans of Bollmann Bildkarten of Braunschweig, the town plans and route maps of R. De Rouck of Brussels, the splendid Austrian walkers' maps (*wanderkarten*) of Kompass Verlag or Freytag and Berndt, the maps and educational atlases of the Kartographische Institut Bertelsmann or the maps of TCI (Touring Club Italiano).

THE UNITED STATES OF AMERICA

In a car-conscious nation such as the USA it is not surprising that the automobile map dominates the cartographic scene, but there is also a thriving commercial mapmaking industry producing a varied range of maps and atlases. Some 120 firms are engaged in this sector and Jon. M. Leverenz separates them into three categories based on the complexity of their production methods and the sophistication and variety of their products (*The American Cartographer*, vol. 1, No. 2 (1974), p. 118). The largest number of firms (about seventy) produce one- or two-colour street maps or tourist maps of local interest which require very little interpretation or generalisation from the original base material used. Little creative work is involved and the product serves the need of people in small urban communities and rural areas. The second category produces a similar range of products but in rather more sophisticated form, the map preparation involving more interpretation and modification of the source materials. Four-colour printing is used and the map users tend to be banks, real estate companies and publishers in larger urban centres. The third category has an extensive range of products varying from globes and wall maps to road maps and general and thematic atlases. Considerable modification of source materials is involved and the full spectrum of operations from research, through compilation and drafting to full-colour printing is required. This sector of the industry has a national and international market and although there are only about fifteen firms in this category they employ almost half the total number of cartographers in the commercial field.

Probably the most familiar American maps to readers throughout the world are the periodic map supplements which appear with the *National Geographic* magazine. These include not only the distinctive reference maps, packed with placenames and verbal information, but rather better examples of cartography in the series of maps on historical themes and the remarkable maps of the ocean floors painted by the distinguished Alpine artist, Heinrich Berann.

ATLAS PRODUCTION IN THE MODERN WORLD

In addition to official national atlases, many commercially produced atlases are available to suit most needs and pockets. The general atlas is one of the most frequently-consulted map sources in modern society. It appears in formats ranging from small pocket-sized atlases to bulky tomes, multi-volume sets, and large loose-leaf atlases which are periodically supplemented by additional sheets on new themes. The requirements of the specialist user are met by atlases devoted to specific themes – history, geology, economics, the Bible, climate, disease, language and so on. The general user, however, needs a modestly-priced reference atlas, and the number which are on the market already means that the atlas publisher has to produce something special in the way of fine colour printing, clarity and overall design in order to attract potential customers. In the higher price range of general atlases are the single-volume Comprehensive Edition of the *Times Atlas of the World* and the Rand McNally *International Atlas*. The latter is unusual in being international in concept, planning, editorial policy and production. Distinguished cartographers and geographers from all over the world contributed to the atlas and leading cartographic establishments in several countries participated in the project. The result is an outstanding example of modern atlas production. A brief summary of its content serves to indicate the wide field of information supplied by an atlas of this quality: first is a section of thematic maps concerned with rainfall, vegetation, industry and so on, together with explanatory text and colour photographs; this section is followed by 'portrait' maps designed to provide an overall impression of physical features throughout the world, including the configuration of the ocean floors; then come the politically-coloured maps which form an important section of any general atlas (how many readers have had their impressions of a country conditioned by the colour chosen for it in their favourite atlas? Mark Twain pointed this out when his characters, Tom Sawyer and Huckleberry Finn, took a trip in a balloon and Huck told Tom that he knew they were over Illinois by the colour. 'Illinois is green . . . I've seen it on the map'). The political maps in this atlas are sensibly presented on an equal-area projection to facilitate comparisons of size. The ensuing physical maps feature a combination of shaded relief and layer tints in a striking colour progression ranging from pale blue-green, through greens, to browns, to yellow and finally to white for the highest land; a final section features plans of important urban areas and maps of strategically-significant regions. The atlas also includes statistical tables, a glossary and an index of some 160,000 names.

Atlases on a more modest scale include the *Atlas Advanced* (Collins–Longman), *Goode's World Atlas* (Rand McNally), *University Atlas* (Philip) and *The Edinburgh World Atlas* (Bartholomew). One of the best-selling general atlases of recent years has been the *Readers' Digest Great World Atlas* which is well suited to home reference use and study.

Space precludes further mention of the multiplicity of available atlases, except to comment on the elegance of the *World Atlas of Wine* and *World Atlas of Food* (Mitchell-Beazley), two specialised atlases of wide appeal which set the highest standards of contemporary book and atlas production, and on the *Times Atlas of World History* (Times Books), a volume which breaks away from the traditional approach to historical atlases. In it, unusual projections, distorted projections, varied orientations, striking colours and bold symbols are used in order to make a bold impact and to give a feeling of dynamism. In mapping the expansion of Islam in the seventh century, for example, the compilers have wisely centred their map on Mecca, and in illustrating the sixteenth century Venetian empire they have orientated the map to the south so that the viewer is presented with a picture of the empire seen, as it were, from Venice itself. In general, illustrators and not cartographers were used to make the maps and their influence is seen in the bold use of surging arrows in striking colours.

THE USE OF MAPS IN PUBLICITY AND ADVERTISING

Aside from the more traditional forms of cartography, there are several fringe cartographic activities which employ maps in unusual ways. In post-war years the advertising executive has come to appreciate the value of maps as a publicity aid and deploys them in an extensive range of product promotions. Maps are naturally more appropriate in some areas than others – in the travel and tourist trades, the food and drink business, or whenever it is useful to emphasise the far-ranging distribution of a product. In other words, they are especially useful when geography is of some significance. On the other hand, many advertisers merely use maps as an attractive element in an overall design, often utilising an early map of decorative character for the purpose.

Whatever the purpose of the advertisement or the type of map used, the promoter has a selection of communications media by which to spread his message – the press, magazines, posters, publicity leaflets and brochures, window displays, television, hoardings and the cinema. A singular feature of mapwork for publicity is that the cartographer in the traditional sense is rarely involved, the schemes of the ideas men being brought to visual life by teams of graphic artists. As was the case in the *Times Atlas of World History* the graphic artist is evidently deemed more capable of producing dynamic, dramatic effects than the cartographer. A second curious point, perhaps with some bearing on the psychology of map users, is that while manufacturers of products for exclusive masculine use frequently feature maps in their publicity promotions, this is seldom true of products used entirely by women! Does this indicate that maps in general have a greater appeal and interest for the male rather than the female sex, or is it simply that advertisers tend to associate maps with masculinity? The question suggests a nice line of research!

Travel and tourist agencies produce an impressive annual array of maps of all types, ranging from conventional topographical maps such as the splendid map of Swiss railways and post coach routes prepared by Kummerley and Frey for Swiss Federal Railways, to

Fig 125 Heraldic map of Devonshire (1972) by T. Alan Keith-Hill. The superbly-executed calligraphic border forms an effective surround to the map and its colourful heraldry.
Courtesy: T. Alan Keith-Hill

accents must be made which stamp it as a subjective picture. Mountain peaks, cliffs, characteristic elevations are emphasised, otherwise not even the highest peak would stand out sufficiently in a broad, panoramic view'. In other words, Berann is saying that a panoramic map is not necessarily absolutely correct in every detail but is designed to provide an instantly recognisable composite picture. Though primarily an artist, Berann takes every advantage of modern technology. He flies over the selected area in a light plane, a topographical map ready for instant consultation, making sketches of features hidden to the earthbound observer and capturing characteristic landscapes with his camera. Back in his studio he keeps files of oblique aerial photographs available for reference. The basic elements of Berann's work, therefore, are maps, sketches, photographs and visual impressions gained from ground and air. He decides

which direction of view will provide the most striking impression of a scene and his panoramas are not always orientated to north. Different impressions are sometimes required to convey the attractions of an area at different seasons for the summer tourist or the winter skier; the former needs the emphasis to be placed on the charms of meadows, lakes, forest and wide sunny valleys, while for the skier it may be necessary to introduce some distortion into the portrayal of slopes in order to emphasise the quality of the facilities for ski-ing. Before leaving Berann, further mention must be made of his relief maps of the ocean floors painted for the *National Geographic* from physiographic diagrams compiled by Bruce Heezen and Marie Tarp. These paintings have aroused universal interest for they lay bare vast panoramas unseen by the human eye, just as Philip Buache tried to do in the eighteenth century with the much less sophisticated resources at his disposal.

This suggestion that the basic aims and objectives of mapmaking have changed little is perhaps a fitting note on which to close. The twentieth century has been one of

Fig 126 **A panoramic map, painted for the 1976 Winter Olympics held in Innsbruck, by the distinguished Alpine artist, Heinrich Berann.**
Courtesy: **Heinrich Berann**

radical change in the technology of cartography, and mapmakers have added new skills to their stock-in-trade; photogrammetry, remote sensing, the use of highly sophisticated survey instruments, the application of automated processes to topographical mapmaking, the use of the computer in processing and transforming data. Cartography is firmly in the grip of technology but happily, as we have seen in the closing pages, a flourishing minority of mapmakers still produce the type of work in which artistry reigns supreme. What does the future hold for cartography? With the establishment of increasing numbers of data banks, the computer will be brought more and more into play. Space technology, allied to remote sensing devices, will assume an increasing role in environmental examination with the emphasis on the conservation of resources, the control of pollution, transportation and urban planning. Will maps in their present form as a printed sheet of paper remain a viable product or will the map user of the future simply call up any map he wishes by computer for viewing on a cathode ray tube?

The only certainty seems to be that cartography will continue to develop excitingly and assume even greater importance in its traditional role of recording and communicating information about the human environment.

SELECT BIBLIOGRAPHY

RECOMMENDED FOR GENERAL READING

DICKINSON, G. C., *Maps and air photographs*, Edward Arnold, London, 1969.

GREENHOOD, D., *Mapping*, University of Chicago Press, Chicago, 1964.

MUEHRCKE, PHILLIP C., *Map use, reading and analysis*, JP Publications, Madison WI, 1978.

ROBINSON, A. H., SALE, RANDALL and MORRISON, JOEL, *Elements of cartography*, 4th ed., John Wiley, New York, 1978.

THROWER, NORMAN J. W., *Maps and man: an examination of cartography in relation to culture and civilisation*, Prentice-Hall Inc., Englewood Cliffs, N.J., 1972.

WOODWARD, DAVID (ed.), *Five centuries of map printing*, University of Chicago Press, Chicago, 1975.

SUGGESTED READING FOR FURTHER INVESTIGATION

1 Maps as a medium of communication

BOARD, C., 'Maps as models' in CHORLEY, R. J. and HAGGETT, P., *Models in geography*, Methuen, London, 1967, pp. 671–725.

FORDHAM, SIR H. G. F., *Maps: their history, characteristics and uses*, Cambridge University Press, Cambridge, 1921.

GUELKE, LEONARD (ed.), *The nature of cartographic communication*, Cartographica Monograph No. 19, Toronto, 1977.

HILL, GILLIAN, *Cartographical curiosities*, British Library, London, 1978.

HODGKISS, A. G., 'The geographer as mapmaker; the role of cartography in recording the man-environment relationship', in Open University Course D204, Section 1 Unit 3, Milton Keynes, 1977.

MUEHRCKE, PHILLIP C., and MUEHRCKE, JULIANA O., 'Maps in literature', *The Geographical Review*, Vol. LXIV No. 3 (1974), pp. 317–38.

ROBINSON, A. H. and PETCHENIK, B. B., *The nature of maps: essays toward understanding maps and mapping*, University of Chicago Press, Chicago, 1976.

2 Fundamentals of mapmaking (i) The development of a cartographic language

CERNY, J. W. and WILSON, JOHN, 'The effect of orientation on the recognition of simple maps', *Canadian Cartographer*, Vol. 13 No. 2, Toronto (1976), pp. 132–38.

CHAMBERLIN, W., *The round earth on flat paper. A description of the map projections used by cartographers*, National Geographical Society, Washington DC, 1947.

MALING, D. H., *Co-ordinate systems and map projections*, George Philip, London, 1973.

STEERS, J. A., *An introduction to the study of map projections*, University of London Press, London, 13th ed., 1962.

3 Fundamentals of mapmaking (ii) The development of a cartographic vocabulary

ALIPRANDI, LAURA and GIORGIO with POMELLA, MASSIMO, *Le Grandi Alpi nella cartografia*, Priuli & Verlucci, Ivrea, 1974.

DAINVILLE, S. J. FRANÇOIS DE, *Le Langage des géographes. Termes, signes, couleurs des cartes anciennes 1500–1800*, Picard, Paris, 1964.

ENGELHART, BEN and BRAND, CHRIS, *Gerardus Mercator: cartographer and writing master*, Werner Renckhoff KG, Duisburg, 1962.

GEORGE, WILMA, *Animals and maps*, Secker & Warburg, London, 1969.

IMHOF, EDUARD, *Kartographische geländedarstellung*, Walter de Gruyter, Berlin, 1965.

KEATES, J. C., 'Symbols and meaning in topographic maps', *International yearbook of cartography*, Vol. 12 (1972), pp. 168–80.

LYNAM, EDWARD, 'Period ornament, writing and symbols on maps', in *The mapmaker's art*, Batchworth Press, London, 1953, pp. 37–54.

OSLEY, A. S., *Mercator. A monograph on the lettering of maps, etc. in the sixteenth-century Netherlands*, Faber & Faber, London, 1969.

ROBINSON, A. H., *The look of maps: an examination of cartographic design*, University of Wisconsin, Madison WI, 1952.

4 Making the map

BOARD, C. and TAYLOR, R. M., 'Perception and maps; human factors in map design and interpretation', Institute of British Geographers, New Series, Vol. 2 No. 1, 1972, pp. 19–36.

CLAIR, COLIN, *A Chronology of Printing*, Cassell, London, 1969.

DILKE, O. A., *The Roman Land Surveyors: An Introduction to the Agrimensores*, David & Charles, Newton Abbot, 1971.

EXPERIMENTAL CARTOGRAPHY UNIT, Royal College of Art, *Automated cartography and planning*, Architectural Press, London, 1969.

HIND, A. M., *A history of engraving and etching from the fifteenth century to the year 1914*, reprinted, Dover Publications, New York, 1963.

HINKS, A. R., *Maps and survey*, Cambridge University Press, 5th ed., Cambridge, 1947.

KEATES, J. S., *Cartographic design and production*, Longman, Harlow, 1973.

LEWIS, JOHN, *Anatomy of printing; the influences of art and history on its design*, Faber & Faber, London, 1970.

PIZZEY, STEPHEN, *Exploring remote sensing*, HMSO, London, 1977.

Precision in mapmaking, Monotype Recorder Vol. 43 No. 1, 1964.

RAISZ, E., *General cartography*, McGraw-Hill, New York, 1948.

RHIND, D., 'Computer-aided cartography', Institute of British Geographers New Series, Vol. 2 No. 1 (1977), pp. 71–96.

RICHESON, A. W., *English land measuring to 1800; instruments and practice*, Society for the History of Technology and the MIT Press, Cambridge, Mass., and London, 1966.

RUDD, ROBERT D., *Remote sensing; a better view*, Duxbury Press, North Scituate, Rhode Island, 1974.

SKELTON, R. A., 'The early map printer and his problems', *Penrose Annual*, Lund-Humphries, Vol. 67 (London, 1964), pp. 171–84.

WARTNABY, J., *Surveying instruments and methods*, HMSO, London, 1968.

WOODWARD, D. (ed.), *Five centuries of map printing*, University of Chicago Press, Chicago, 1975.

5 The evolving world map

BAGROW, L., *History of cartography* (ed. R. A. Skelton), C. A. Watts & Co., London, 1964.

BEAZLEY, C. RAYMOND, *The dawn of modern geography*, Vol. 2, Peter Smith, New York, 1949.

BRICKER, C. with TOOLEY, R. V., *A history of cartography: 2,000 years of mapmaking*, Thames & Hudson, London, 1969.

BROWN, LLOYD A., *The story of maps*, Little, Brown, Boston, Mass., 1949.

CRONE, G. R., *Maps and their makers: an introduction to the history of cartography*, revised edition, Dawson, Folkestone, 1978.

CRONE, G. R., *The Hereford world map*, Royal Geographical Society, London, 1948.

DESTOMBES, M. (ed.), *Mappemondes AD 1200—1500*, Imago Mundi Supplements, Vol. 4, N. Israel, Amsterdam, 1964.

GROSJEAN, G. and KINAEUR, R., *Kartenkunst und kartentechnik von altertum bis zum barock*, Hallwag, Bern and Stuttgart, 1970.

JOHNSON, H. B., *Carta marina: world geography in Strasburg, 1525*, University of Minnesota Press, Minneapolis, 1963.

LISTER, R., *How to identify old maps and globes*, Bell, London, 1965.

LISTER, R., *Antique maps and their cartographers*, Bell, London, 1970.

NORDENSKIÖLD, A. E., *Facsimile atlas to the early history of cartography*, reprint by Dover Publications Inc., New York, 1973.

SANZ, CARLOS, *Ciento noventa mapas antiquas del mundo de los siglos 1 al XVIII*, Real Sociedad Geografica, Madrid, 1970.

SKELTON, R. A., *Explorers' maps: chapters in the cartographical record of geographical discovery*, Spring Books, London, 1970.

STEVENS, Henry N., *Ptolemy's geography*, reprint by TOT, Amsterdam, n.d.

TOOLEY, R. V., *Maps and mapmakers*, 4th ed., Batsford, London, 1970.

VRIJ, MARIJKEDE, *The world on paper. A descriptive catalogue of cartographical material published in Amsterdam during the seventeenth century*, Amsterdam's Historisch Museum, 1967.

YONGE, ENA L., *A catalogue of early globes*, American Geographical Society Library Series No. 6, New York, 1968.

6 Regional maps

BARTHOLOMEW, JOHN, & SON LTD, *The mapping of Scotland*, Bartholomew, Edinburgh, 1971.

BRAMSEN, BO, *Gamle danmarkskort 1570–1770*, Politikens Forlag, Copenhagen, 1952.

BRATT, EINAR, *En krönika om kartor över Sverige*, GLA, Stockholm, 1958.

BRITISH MUSEUM, *The mapping of the British Isles, thirteenth to nineteenth centuries*, Catalogue of an exhibition in the King's Library, British Museum, London, 1964.

BUCZEK, KAROL, *The history of Polish cartography from the fifteenth to the eighteenth century*, Ossolineum, Warsaw, 1966.

CHUBB, THOMAS, *The printed maps in the atlases of Great Britain and Ireland: a bibliography, 1579–1870*, reprint by Dawson, London, 1966.

CORTESÃO, ARMANDO and TEIXIERA DA MOTA, ARELINO, *Portugaliae monumenta cartographica*, Lisbon, 1960.

CLOSE, SIR CHARLES, *The early years of the Ordnance Survey*, David & Charles reprint, Newton Abbot, 1969.

CRONE, G. R., *Early maps of the British Isles*, introduction to Royal Geographical Society reproductions of early maps, No. VII, Royal Geographical Society, London, 1961.

CUMMING, W. P., *Cartography of colonial America*, University of South Carolina, Columbia, 1961.

FITE, EMERSON D. and FREEMAN, ARCHIBALD, *A book of old maps delineating American history*, Dover Publications Inc., New York, 1969.

GROSJEAN, GEORGES, *500 jahre Schweizer landkarten*, Orrell Fussli Verlag, 1971.

KOEMAN, DR IR. C., *Joan Blaeu and his Grand Atlas*, Philip, London, 1970.

KOEMAN, DR IR. C., *Atlantis Neerlandici*, TOT, Amsterdam, 1969.

LE GEAR, CLARA EGLI (compiler), *United States atlases: a list of national, state, county, city and regional atlases in the Library of Congress*, Vol. 5, 1958, and Vol. 6, 1963, Government Printing Office, Washington.

LYNAM, E., 'Saxton's atlas of England and Wales' in *The mapmaker's art*, Batchworth Press, London, 1953, pp. 79–90.

LYNAM, E., *British maps and mapmakers*, Collins, London, 1947.

NANBA, M. with MUROGA, N. and UNNO, K., *Old maps in Japan*, Sogensha Inc., Osaka, 1973.

NEEDHAM, JOSEPH, 'Geography and cartography' in *Science and civilisation in China*, Vol. 3 (Cambridge University Press, 1959), pp. 497–590.

NORTH, F. J., *The map of Wales before 1600 AD*, National Museum of Wales and University of Wales, Cardiff, 1935.

PHILLIPS, PHILIP LEE, *A list of maps of America in the Library of Congress*, TOT, Amsterdam reprint, 1967.

RISTOW, WALTER W., *Guide to the history of cartography: an annotated list of references on the history of maps and mapmaking*, Library of Congress, Washington, 1973.

RISTOW, WALTER W. (compiler), *A La Carte: selected papers on maps and atlases*, Library of Congress, Washington, 1972.

ROYAL SCOTTISH GEOGRAPHICAL SOCIETY, *The early maps of Scotland to 1850*, Royal Scottish Geographical Society, Edinburgh, 1973.

SKELTON, R. A., *County atlases of the British Isles 1579–1703*, Carta Press, London, 1970; reprint Dawson, Folkestone, 1978.

TOOLEY, R. V., *Collector's guide to maps of the African continent*, Carta Press, London, 1969.

7 Nautical charts

DAY, ARCHIBALD, *The Admiralty Hydrographic Service*, HMSO, London, 1967.

EDGELL, SIR JOHN, *Sea surveys: Britain's contribution to hydrography*, HMSO, London, 1965.

HOUSE, DEREK and SANDERSON, M., *The sea chart*, David & Charles, Newton Abbot, 1973.

NORDENSKIÖLD, A. E., *Periplus: an essay on the early history of charts and sailing directions*, Stockholm, 1897.

PARRY, J. H., *An illustrated history of men, ships and the sea in the fifteenth and sixteenth centuries*, Dial Press, New York, 1974.

RITCHIE, G. S., *The Admiralty chart. British naval hydrography in the nineteenth century*, Hollis & Carter, London, 1967.

ROBINSON, A. H. W., *Marine cartography in Britain. A history of the sea chart to 1855*, Leicester University Press, Leicester, 1962.

TAYLOR, E. G. R., *The haven-finding art*, Hollis & Carter, London, 1956.

8 Route maps

CAMPBELL, T., 'The woodcut map considered as a physical object: a new look at Erhard Etzlaub's *Rom Weg* map of c 1500', *Imago Mundi* 30, Second Series Vol. 4, Lympne, 1978.

DISLEY, J., *Orienteering*, Stackpole Books, Harrisburg, Penn., 1973.

FORDHAM, SIR H. G., *John Ogilby (1600–1676), his Britannia, and the British itineraries of the eighteenth century*, Oxford University Press, Oxford, 1925.

FORDHAM, SIR H. G., *The road-books and itineraries of Great Britain, 1570–1850*, Cambridge University Press, Cambridge, 1924.

FORDHAM, SIR H. G., *The road-books and itineraries of Ireland*, J. Falconer, Dublin, 1923.

HARLEY, J. B., 'Maps of communication', in *Maps for the*

local historian, Standing Conference for Local History, London, 1972, pp. 40–50.

MODELSKI, ANDREW M. (compiler), *Railroad maps of the United States*, Library of Congress, Washington, 1975.

RISTOW, WALTER W., *Aviation cartography: a historico-bibliographic study of aeronautical charts*, Library of Congress, Washington, 1960.

RISTOW, WALTER W. (ed.), *Christopher Collis: a survey of the roads of the United States of America, 1789*, Harvard University Press, Cambridge, Mass., 1961.

RISTOW, WALTER W., 'A half century of oil-company road maps', reprinted from *Surveying and mapping*, Vol. XXIV, No. 4 (Washington, 1964), pp. 617–37.

9 Town plans and views

BONSOR, K. J. and NICHOLS, H., *Printed maps and plans of Leeds, 1711–1900*, Thoresby Society, Leeds, 1958.

DARLINGTON, IDA and HOWGEGO, J., *Printed maps of London circa 1553–1850*, Philip, London, 1964; 2nd edition by J. Howgego, Dawson, Folkestone, 1978.

FORDHAM, ANGELA, *Town plans of the British Isles*, Map Collectors' Series 22, Map Collectors' Circle, London, 1965.

GLANVILLE, PHILIPPA, *London in maps*, Connoisseur Books Division, 1972.

HÉBERT, JOHN R., *Panoramic maps of Anglo-American cities*, Library of Congress, Washington, 1974.

HYDE, RALPH, *Printed maps of Victorian London: 1851–1900*, Dawson, Folkestone, 1975.

MAP COLLECTORS' CIRCLE, *North American city plans*, Map Collectors' Series 20, Map Collectors' Circle, London, 1965.

WELCH, E., *Southampton maps from Elizabethan times*, Southampton City Council, 1964.

10 Thematic Maps

DAVIS, PETER, *Data description and presentation*, Oxford University Press, Oxford, 1974.

DICKINSON, G. C., *Statistical mapping and the presentation of statistics*, Edward Arnold, London, 2nd edition, 1974.

ENGELMAN, G., 'Der Physikalischer Atlas der Heinrich Berghaus and Alexander Keith Johnston's Physical Atlas', *Petermann's geographische mitteilungen*, Vol. 108 (1964), pp. 133–49.

FRIIS, H. R., 'Statistical cartography in the United States prior to 1870 and the role of Joseph C. G. Kennedy and the US Census Office', *The American Cartographer*, Vol. 1 No. 2 (Washington, 1974), pp. 131–57.

GREGORY, S., *Statistical methods and the geographer*, Longman, London, 1963.

HODGKISS, A. G., *Maps for books and theses*, David and Charles, Newton Abbot, 1970.

LAWRENCE, G. R. P., *Cartographic methods*, Methuen, London, 1971.

LIEBENBERG, ELRI, 'Symap: its uses and abuses', *Cartographic journal*, Vol. 13 No. 1 (June, 1956), pp. 26–35.

MONKHOUSE, F. J. and WILKINSON, H. R., *Maps and diagrams*, 3rd edition, Methuen, London, 1971.

RAISZ, E., *Principles of Cartography*, McGraw-Hill, New York, 1962.

ROBINSON, A. H., 'The 1837 maps of Henry Dury Harness', *Geographical Journal*, Vol. 121 (1955), pp. 440–50.

ROBINSON, A. H., 'The thematic maps of Charles Joseph Minard', *Imago Mundi*, Vol. 21 (1967), pp. 95–108.

ROBINSON, A. H. and WALLIS, HELEN, 'Humboldt's map of isothermal lines: a milestone in thematic cartography', *Cartographic Journal*, Vol. 4 (1967), pp. 119–23.

TRURAN, H. C., *A practical guide to statistical maps and diagrams*, Heinemann, London, 1978.

11 Official mapmaking today

ALEXANDER, G. L., *Guide to atlases: an international listing of atlases published since 1950*, Scarecrow Press, Metuchen, N. J., 1971.

HARLEY, J. B., *The historian's guide to Ordnance Survey maps*, National Council of Social Service, London, 1964.

HARLEY, J. B., *Ordnance Survey maps: a descriptive manual*, Ordnance Survey, Southampton, 1975.

LAYNG, T. E., 'Highlights in the mapping of Canada', reprinted from *Canadian Library*, Ottawa, 1960.

LOCK, C. B. MURIEL, *Modern maps and atlases*, Clive Bingley, London, 1969.

MONOTYPE RECORDER, *Precision in mapmaking*, Monotype Corporation Ltd, London, 1964.

OLSON, EVERETT C. and WHITMARSH, AGNES, *Foreign maps*, Harper, New York, 1944.

ROBINSON, A. H., MORRISON, JOEL L. and MUEHRCKE, PHILLIP C., 'Cartography, 1950–2000', Institute of British Geographers, trans. New Series, Vol. 2 No. 1 (1977), pp. 3–12.

WINCH, KENNETH L. (ed.), *International maps and atlases in print*, Bowker, London, 1974.

12 Modern commercial and private cartography

AGER, JOHN, 'Maps and propaganda', Society of University Cartographers, *SUC Bulletin*, Vol. 11 No. 1 (Portsmouth, 1977), pp. 1–14.

BERANN, H. C. with GRAEF, H. A., *Die Alpen im panorama*, Verlag-Weidlich, Frankfurt, 1966.

Braunschweig Städtisches Museum, *Hermann Bollmann*, Braunschweig, 1966.

GARFIELD, T., 'The panorama and reliefkarte of Heinrich Berann', Society of University Cartographers, *SUC Bulletin*, Vol. 4 No. 2, Liverpool, 1970.

HODGKISS, A. G., 'The bildkarten of Hermann Bollmann', *The Canadian Cartographer*, Vol. 10 No. 2, Toronto, 1965.

KEITH-HILL, T. ALAN, 'Modern heraldic county maps', Society of University Cartographers, *SUC Bulletin*, Vol. 8 No. 1, Liverpool, 1973.

LEVERENZ, JON M., 'The private cartographic industry in the United States; its staff and educational requirements', *The American Cartographer*, Vol. 1 No. 2 (Washington, 1974), pp. 117–23.

MacDERMOTT, PAUL D., 'Cartography in advertising', *The Canadian Cartographer*, Vol. 6 No. 2, Toronto, 1969.

QUAM, L. O., 'The use of maps in propaganda', *Journal of Geography*, Vol. 42 (1943), pp. 21–32.

WAINWRIGHT, A., *Fellwanderer: the story behind the Guidebooks*, Westmorland Gazette, Kendal, 1966.

INDEX

Numbers of illustrations are in italics

Adams, John 123, 124
Administrative maps 172
Advertising maps 188, 189, 191, *119, 120, 121*
Aerial photography 55–8
Aeronautical charts 15, 129–32
Agnese, Battista 108
Agrimensor 155
Agrippa, Marcus Vipsanius 73, 119
Airey, John 129
Akin, William B. 125
Alingham, William 11
American city plans 137, 147, *96, 100*
American county atlases 101
Anaximander's world map 72
Ancelin, Pierre 156
Animals on maps 47
Apian, Peter 91
Apian, Philip 44, 53, 63, 64, 92
Ashley, Anthony 109
Atlas production 187
Australian national map series 179
Automated cartography 69
Automobile maps 125, 187

Babylonian clay tablets 25, 27, 29, 39, 44, 61, 86, 155, *40, 54*
Balloon views 148, *97*
Bartholomew, John & Son Ltd 42, 184
Baylie, Charles 43, 159
Beaufort, Thomas 115
Beatus, *Commentary on the Apocalypse* 75
Beechey, F. W. 159
Behaim, Martin 80
Bellin, J. N. 112, 147
Belser Verlag 187
Berann, Heinrich 29, 187, 194, 196, 197, *126*
Berey, Nicholas 66
Berghaus, Heinrich 161, 162, 105
Berlin, Atlas of 102, 151
Berlinghieri, Francesco 40, 90
Bertelsmann Kartographische Institut 187
Bianco, Andrea 79
Bird's eye view 45, *85*
Blackmore, Nathaniel 156
Blaeu, Joan 30, 49, 95, 141, *92*
Blaeu, W. J. 40, 46, 95, 109
Board of Health plans 149, 150
Bodleian, or Gough map 44, 89, 121, *57*
Bollmann, Hermann 151, 153, 187, *99, 100*
Bonne, Rigobert 83
Bonner, John 147
Booth, Charles, poverty maps 167
Bordone, Benedetto 106, 133, 137, *37, 85*
Bowen, Emanuel 124
Bowes, William 16
Bradford, William 147
Bradshaw, George 128, 129, *82*
Braun, Georg and Franz Hogenberg, *Civitates Orbis Terrarum* 137, 139, 145, *90*

Brétez, Louis 145, *94*
Bruinss, Pieter 156
Bryant, Andrew 96, 99
Buache, Philippe 40, 99, 156, 157, 158
Buck, Samuel and Nathaniel 147, 148
Buondelmonte, Christopher 106, 133
Burdett, P. P. 127

Cadastral maps 19, 55, 71, 155
Calligraphy 48–51, *27, 85*
Camden, William, *Britannia* 123
Camera-ready drawing 21
Camocio, Gian Francesco 107
Canadian cartography 101, 102, 179, 180, *114*
Canadian national map series 179, 180
Carte Pisane 25, 104
Cartesian co-ordinates 30, *16*
Cartogram 86
Cary, John 58, 96, 121, 124, 127, 168
Cassini family 53, 112, 170, *31*
Cassini, Jean Dominique 112
Catalan Atlas 104, 106, *69*
Catalan chartmakers 104
Celtis, Konrad 119
Centuriation 30, *16*, 155
Cerography 68
Chang Heng 30, 78
Chapin, R. M. 194
Cheng Ho 108
Child, Heather 191
Chinese maps 39, 52, 53, 76–8, 80, 86–7, 108, 127
    panoramic maps 127
    Phei Hsui 78
    regional maps 86, 87
    sailing charts 108
    on silk 87
    waterways maps 127
Chorobates, levelling instrument 52
Chorography 89
Christian Aid 38, 194
Chu, Ssu-Pén 98
Church as settlement symbol 44, *24*
Clarendon Press 184
Classification of maps 19
Clavus, Claudius 90
Climatic zonal world maps 75, *46*
Coastlines 43, 44, *23*
Cole, Humphrey 53
Collins, Greenvile, *Great Britain's Coasting Pilot* 109, 112, *73*
Colom, Arnold 109
Colom, Jacob Aertz 109
Colours 66, 193
Colton, Joseph Hutchins 129
Commercial mapmaking 184–8
Communication in cartography 11–24, *9, 10*
Compilation 58–61
Computer mapping 69, *39*
Contarini, G. M. 82, *51*

Continental Trailways bus system *101*
Contours 40, 42, 160, 161
Cook, Captain James 115
Coronelli, P. Vincenzo Maria *12*
Cosmographiae 90, 91
Cottonian, or Anglo-Saxon map 75, 76, *47*
County atlases, American 101
Cresques, Abraham 106, *69*
Cresques, Jafudo 106
Cross-staff 52
Cruquius, N. S. 40, 156, *102*
Cunningham, William 139

Dalrymple, Alexander 115
Danish national map series 175
Dasymetric technique 164
Data, absolute or derived 154
Data banks 69
Datum 42, 154
Dawson, Lt R. K. 148, 149
de Beaurain, S^r. *93*
de Brahm, William Gerard 114
de Bry, Theodore 109, *66*
Defoe, Daniel 18
de la Cosa, Juan 80, 82, 83, *50*, *53*
de l'Isle, N. 99
Denham, Captain H. M. *76*
Density map 21
Department of Energy, Mines and Resources, Canada 179
Department of Lands and Survey, New Zealand 179
De Rouck – Belgian maps 187
Des Barres, J. F. W. 49, 114, 115, 128, *75*
Desceliers, Pierre 108, *34*
Dezauche, J. A. 99
Diagram – London designers 15
Dicey & Co. 66
Dieppe school of chartmaking 108, *4*, *34*
Digges, Leonard 53
Digges, Thomas 159
Digitizing tables 69
Dioptra 52
Directorate of Overseas Survey (DOS) *113*
Disease, maps of 164
Dissected maps 18, *6*
Distribution of maps 21
Doncker, Hendrick 109
Donn, Benjamin 96, *65*
Dry, Camille N. 147, *96*
Dudley, Sir Robert, *Dell' Arcano del Mare* 108
Dupain-Triel, J. L., introduction of contours to land maps 40, 160
Duryea, Frank 125
Dutch cartography 95, 109, 174
du Val, Pierre 18

Earth Resources Observation Systems (ERCS) 58
Eastern European cartography 170, 179
Ebstorf map 61, 104
Educational cartography 15, 16, 185
Educational games, maps in 16, *5*, *6*
Egyptian cartography 52, 72, 155
Electronics communications theory 22, 23
English county maps 95–9, *11*, *62*, *63*, *64*, *65*
English Tourist Board 189
Engraved maps 19, 49, 59, 65, 66, *38*
Eratosthenes, reference system 72, 78, *41*
Eskimo maps 25, 71
Esselte map service, Sweden 175
Etching 65
Etymologiae, of Bishop Isidore of Seville 63
Etzlaub, Erhard 121

Eyes, John 112

Faden, William 147, 170
Fairey Surveys 184
Falk Verlag 186
Fearon, Samuel 112
Finé, Oronce 93
Flora and fauna 46, 47
Flow-line technique 154, 164, *106*
Food from France, 'map models' 191
Forlani, Paolo 108
Form lines round coasts 44
Freytag-Berndt Verlag, *Touristenkarten* 187
Frisius, Gemma, system of triangulation 53, 112
Fryer, David L. 184, *118*
Fyffe Group, use of 'map models' 191, *121*

Gardner, Thomas 124
Gastaldi, G. 44, 92
General-purpose reference maps 19
Generalisation 12
Generalstabens Litografiska Anstalt (GLA) 175
Geo Centre of Stuttgart 169, 186
Geodaetisk Institut 175
Geodetic survey 52
Geographia Ltd 184
Geological maps 168
George, Wilma 47
Germanus, Nicolaus 80, 90, *49*
Germany, national map series 174, 175
Gill, F. Macdonald 129, 191, *2*
Goode's Homolosine Equal-Area Projection 38
Goos, Pieter 109
Gough or Bodleian map 44, 89, 121, *59*
Gousha, H. M. 125
Graduated symbols 164
Grammar of cartography 25–38, *4*, *14*, *16*, *17*, *18*, *19*
Graph plotter *39*
Graphical presentation techniques 164
Graticule 30
Greek mapmaking 52, 72, 73
Green, William 147
Greenwood, Christopher 96, 99
Greyhound Bus Company 154
Grids 30, 31, 32, *16*
Groma 52
Gubbio, plan of 143, *92*
Guettard, Jean-Etienne 157
Gunter, Edmund, measuring chain 55

Hachuring 40
Hadley's octant 114
Halley, Edmond 19, 159, *103*
Hand colouring 66
Harness, Henry D. 164, *106*
Harrison, John, marine chronometers 112
Harrison, Richard Edes 194
Hecataeus 72
Heezen, Bruce 197
Henry the Navigator (Prince Henrique of Portugal) 107
Hereford Map 61, 104, 106, 120, *48*
Hero of Alexander 52
Hills and mountains 39–43, *20*, *21*, *22*
Hoefnagel, Georg 139
Hogenberg, Remigius *62*
Holland, Samuel 115
Homann, J. B. 99, 143
Homem, Diogo 107, 108, *35*
Homem, Lopo 107, 108
Hondius, Jodocus 44, 109

Hong Kong, Crown Lands and Survey Office *111*
Horwood, Richard 147, *95*
Hume, Reverend A. 167, *107*
Humorous maps 18
Hunting Surveys Ltd 184
Hurd, Thomas 115
Hutchins, Thomas 101
Hydrographic Department of the Admiralty 115

al-Idrisi 74, *42, 43*
Imhof, Dr Eduard 42, *22*
India, cartographic tradition 179
Information processing 58–60
Infra-red photography 58
Intaglio printing 65
International Civil Aviation Organisation (ICAO) 170
International Commission for Aeronautical Charts 131
International Hydrographic Bureau 115
International Map of the World (IMW) 169, 170
Interrupted projections 38
Isidore, Bishop of Seville 63, 74
Islamic cartography 73, 74, 88, *42, 43*
Isogonic lines 19, 159, *103*
Isolario, or Books of Islands 106, 107, *37, 70*
Isolines 159, *39, 103*
Isopleths 164
Isotherms 161
al-Istakhri 74
Istituto Geografico Militare 47, 175
Italian chartmakers 104
Italic hand 49
Italy, national map series 175
Itineraries 121

Jaillot, Alexis Hubert 112, 143
James, Sir Henry 67
Jansson, J. 40, 49, 95, 109, *66*
Jefferson, Thomas 101
Jefferys, Thomas 147
Jenkinson, Anthony *26*
Johnston, Alexander Keith 161, 162
Jomard, Edme-François 67
JRO Verlag 186
Julien, R. J. *27*
Justus Perthes Geographische Anstalt 187

Keith-Hill, T. Alan 194, 195, *125*
Key, legend or reference table 44
al-Khwafizimi 73
Kitchin, Thomas 100, 147
Kummerley and Frey 125, 188

Lafreri Atlases 93
LANDSAT 1 *32*
Land-use maps 154, 167, *108*
Language of cartography 25–38
Languedoc Canal map 127
Larcom, Thomas 164
Latini, Brunetto *45*
Laurie, John 127
Laurie and Whittle 125, *6*
Lautensack, Hans 136, *88*
Lea, Philip 124
Leake, John, survey of London 145
Leescaert 104
Legend, *see* Key
Lehmann, G. J. – use of hachuring 40
Leigh, Dr Edwin 167
Leland, John 121
le Testu, Guillaume 108

Lettering 48–51, *29*
Letterpress printing process 28
Levelling 52
Leybourne, William 145
Limitations of maps 11–12
Line illustration 21
Line printer 69
Linschoten, J. H., *Itinerario* 29
Lister, Martin 168
Literature, maps in 19, *7*
Lithography 21, 67–9
Local Government Reorganization Act, 1972 172
London, surveys and plans 139, 145, 147, *90, 95, 97*
London Underground map 108, 129
Loxodrome 108
Lucas, Claude 145, *94*

Macaulay, Zachary 129
Mackenzie, Murdoch 112, *74*
Macrobius, world maps 75, *46*
Madaba mosaic 55, 87, 88
Magini, G. A. 92
Magnetic compass 53
Magnus, Olaus, map of Scandinavia 46, 64
Mahoney, Dorothy 191
Mair's Geographical Publishing House 186
Manuscript maps 19
Map compilation 59–61
Map model technique 191, *121*
Map user 11, 22, 23
Mappae mundi 19, 40, 44, 61, 78–80, 106, *48*
Marine chronometer 112
Mariner's compass 103
Marinus of Tyre 72
Marshall Islanders, 'stick charts' 25, 71
Marsigli, L. F. 156
Mauro, Fra, world map 78, 79, 80
Maury, Matthew Fontaine 167
Medieval maps 74–80, *44, 45, 46, 47, 48, 49, 50, 51, 52, 53*
Mercator, Gerhard 16, 30, 35, 44, 49, 58, 85, 93, 108, 109, 112, 118, *53*
Mercator, Michael *18*
Mercator projection 35, 85, 108, 109, 112, 118
Merian, Caspar 141
Merian, Matthäus 139, 141, 145
Mersey Docks and Harbour Company *77*
Mexican road maps 121
Michelin, André 187
Michelin guides and maps series 187
Michelin Tyre Company 125
Microwave Radiometer 58
Military maps 15, 143, *93*
Miller, Konrad *120*
Milne, Thomas, land-use map 167, *108*
Minard, Charles Joseph 164
Mitchell, John 49, 100, *29, 67*
Moedebeck, Hermann 129, 130
Moll, Hermann 99
Mollweide projection 33
Mongolian style 88
Mooney, John 43, 159
Morden, Robert 18, 123, *79*
More, Sir Thomas 18, *7*
Morgan, William 145
Morris, Lewis 112
Morse, Sydney Edwards, inventor of cerography 68
Mortier, Pierre 112, 143
Mosaic 55, 87, 88
Motoring maps 15, 125
Moule, Thomas 128

Münster, Sebastian 40, 43, 46, 64, 90, 135, *25*, *28*, *58*, *87*
Mural maps *1*

Napier Publications *124*
National Geographic magazine 194
National Grid 30
National Savings Movement *123*
Nautical charts 15, 103–18, *69*, *72*, *73*, *74*
Netherlands, official map series 174
New Zealand, official map series 179, *177*
Newton, Sir Isaac 53
NOAA 2 Satellite image *33*
'Noise' in cartography 23, 24, 29, *10*, *11*
Nolin, J. B. 127
Norden, John 95, 139, *79*
Nordenskiöld, A. E., *Facsimile-Atlas* 68, *19*
Norges Geografiske Oppmåling 175, 176
Norwegian national map series 175, 176
Nowell, Laurence 95
Nunes, Pedro 108
Nuremberg Chronicle 135, 136, *86*

Oekumene, known world of ancient civilisations 72
Offset printing 69
Ogilby, John 95, 121, 124, 145, *80*
Oil company maps 191
Old English Mile 121
Olives, Bartolomeo 108, *71*
Ontario, *Economic Atlas* 102
Ordnance Survey of Britain 15, 27, 40, 47, 51, 53, 65, 67, 96, 125, 149, 171, 172, *98*, *109*
Orienation 27, 28, *4*, *14*
Orienteering 15, 16
Ornamentation 24, *11*
Ortelius, Abraham 58, 66, 85, 93, 119, 156, *3*, *26*
Outshore hatching 43
Owen, John 124

Panoramic maps and plans 29, 127, 147, 148, 187, 194, 196, 197, *96*, *126*
   of American towns 147, *96*
   by Heinrich Berann 29, 187, 194, 196, 197, *126*
   of British towns 147, 148
   of Chinese rivers and lakes 127
Packe, Christopher 42, 43, 159, *104*
Paper 61
Papyrus 61
Parchment 61
Paris, Matthew 44, 88, 89, 120, 121, 127, *56*
Paterson, Daniel, road-book 124
Penck, Albrecht 169
Pené, Charles 172
Pepys, Samuel 109
Periplus, or coastal pilot 103
Petermann, Augustus 164
Peters, Arno, projection 38
Peutinger Table 73, 108, 119, 120, *78*
Peyto Glacier map *114*
Pharos of Alexandria 103
Phei Hsui 78
Philatelic cartography *1*
Philip, George & Son 184
Photogrammetry, science of 55
Photo-lithography 21
Photomap 58
Picard, Jean, survey of France by triangulation 53
Pie graph, introduction by Minard 164
Pirckheimer, Bidibaldo *59*
Pisan sea chart, or *Carta Pisane* 25, 104
Plane survey 52

Plane table 53
Plantin, Christopher, printing house 109
Playing card maps 16, *5*
Plot, Robert 27
Polimetrum 53
Political cartoon maps 18, *8*
Polo, Marco 79, 106, *69*
Popple, Henry 100
Portolan charts 25, 29, 88, 104–6, *35*
Portolano or navigational guide 103, 104
Priestley, Joseph 128
Prime Meridian 30, *15*
Projections 31–8, *17*, *18*, *19*
Propaganda maps 15, 191, 193, *122*
Ptolemy, Claudius 30, 43, 48, 64, 65, 78, 79, 80, 87, 90, 92, *19*, *21*, *49*, *58*, *59*
Public Health Act, 1848 149
Publicity maps 188–91, *119*, *120*, *121*

Qualitative maps 19, 154, *101*
Quantitative maps 19, 154, 156–67, *102*, *103*, *104*, *106*
Quartermaster General's Office, lithographic press 67

Radar 58
Railway maps 128, 129, *83*, *84*
Raisz, Erwin 55
Ramsden, Jesse, great theodolite 53
Ramusio, G. B. 29, 137, *14*, *60*, *89*
Rand McNally 125, 188
Range, township and section method 30, *16*
Ravn, Niels 164
Redmayne, W. *5*
Reference grids 30–2, 72, *16*
Reference table, *see* Key
Regional maps 86–102, *54*
Reich, Gregorius 53
Reinel, Jorge 107, 108
Reinel, Pedro 107, 108
Relief depiction 39–43, *20*, *21*, *22*
Remote sensing 55–8, *32*, *33*
Renaissance mapmaking 80
Rennell, Major James 179
Representative Fraction (R.F.) 27
Reproduction of maps 21, 61–70
Revision of maps 64, 172
Ribeiro, Diego, world map of 1529 83, 107, *52*, *53*
Richard of Haldingham 78
Ritter, Karl 161
Road-books 95, 121, 123, 124, *80*, *81*
Road maps 119–26, 184, 186
Rocque, John 145
Rolevinck, W. 134
Roman cartography 30, 44, 73, 87, 119, 120, *78*
   Agrippa world map 73
   Peutinger Table 73, 119, 120, *78*
   regional maps 87
Romweg route map 121
Rose, Frederick 18
Rotz, Jean 108, *4*
Route maps 119–32, *78*, *80*, *82*, *83*, *84*
Routier 104
Rudel, Ira W., offset printing press 69
Rutlinger, Johannes 109
Rutter 104
Ruysch, Johan *19*
Ryther, Augustine 109

Santarem, Manuel Francisco de Barros, Viscomte de, facsimile atlas, 67, *50*
Satellite photography 55, *32*, *33*

Satirical maps 18
Savoy, Duchy of, cadastral survey 156
Sayer, R. and J. Bennett 124
Saxton, Christopher 30, 44, 53, 95, 127, *11*, *62*
Scale 12, 25, 27, 60, 61, *12*
  transformation of 60, 61
Schedel, Hartmann, *Nuremberg Chronicle* 134, 135, *86*
Schlecht, Richard 194
Scribing 69
Sea-bed, maps of configuration of 40, *102*
Seas, coastlines and rivers 43, 44, *23*
Seckford, Thomas 95
Seller, John 30
Senefelder, Alois 67
Senex, John 99, 124
Sequarus, Johannes Metellus 121
Seutter, Matthew 99
Sextant 104
Sgrooten, Christiaan 93
Shapter, Thomas, medical maps 164
Sheldon tapestry maps 61, *36*
Sian, Chinese maps on stone, AD1137 88
Sideways Looking Airborne Radar (SLAR) 58
Silk, Chinese maps made on 87
Simplification 12
Skelton, John 189
Skinner, Andrew and George Taylor, road-books 124, *81*
Smith, Charles 96
Smith, Captain John, map of Virginia *68*
Smith, William 95, 121, 139, *63*
Smith, William, 19th-century geological maps 168
Snow, C. P. 169
Snow, Dr John, disease maps 164
Social conditions, maps of 167, *107*
Society of Scribes and Illuminators 194
Soil maps 167
Sonetti, Bartolomeo dalli 106
Space photography 55–8, *32*, *33*
Special-purpose maps 19, 154–68
Speed, John 27, 66, 95, 139, *13*, *64*, *91*
Spilsbury, John, maker of dissected maps 18
Spot heights 42, 159, 160
Ssu-Pen, Chu, map of China 88
Stamps, maps on *1*
Standard parallel 35
Statute mile 121
Steel engraving 66
Stereographic projection 33, *18*
Stevenson, Robert Louis 18
Stipple 43
Strachey, J. 168
Strip map technique 121, *80*, *81*
Surveying 52–5, *30*
Swan, John & Co. *117*
Sweden, national map series 175
Switzerland, national map series 174, *110*
Sylvanus, Bernardus 64
Symbols 39, 43–7, 104, 116, 127, 129, 172, 174, 194, *20*, *21*, *22*, *23*, *24*
Symonson, Philip, map of Kent 44

Tapestry maps 61, *36*
Taylor, George and Andrew Skinner, road-books 124, *81*
Taylor, Isaac 147
Terwoort, Leonart 95, *11*
Tharp, Marie 197
Thematic maps 19, 21, 60, 69, 154–68, 185, *102*, *103*, *104*, *105*, *106*, *108*
  Ancient World 155, 156
  of Heinrich Berghaus 161, *105*

disease maps 164
geological maps 168
of Edmond Halley 159, *103*
of Henry D. Harness 164, *106*
of von Humboldt 161
of Alexander Keith Johnston 161
land-use and soil maps 167, *108*
of Charles Joseph Minard 164
objectives of 60
of Augustus Petermann 164
physiographical maps 159, 160, 161, *104*
produced by computer aid 69
qualitative maps 19, 154
quantitative maps 19, 154
social conditions 167, *107*
statistical mapping in 19th-century USA 167
submarine contour maps 156–8, *102*
topographical base 60
Theodolite 43
Times Atlases 188
Tithe Commutation Act 156
Tithe plans 148
T-O maps 74, 75, *44*, *45*
Tolkien, J. R. R. 18
Topographical maps 15, 21, 39, 169–83, *109*, *110*, *111*, *112*, *113*, *114*, *115*, *116*
Tordesillas, Treaty of 30, *82*
Touring Club Italiano 187
Tourist maps 15
Tournachon, Gaspard Felix, *or* Nadar 55
Town plans 138–53
  American city plans 147, 148, *96*
  Balloon views 148, *92*
  Blaeu town books 141, 92
  Board of Health plans 149, 150
  Bollmann '*bildkarten*' 151, 152, *99*, *100*
  British plans and views 139, 148, 149, *91*
  in Isolario 133, 134, *85*
  London plans 145, 147, *90*, *95*, *97*
  Matthäus Merian 139, 141
  military plans 143, *93*
  modern plans 151
  Ordnance Survey urban plans 149, *98*
  woodcut plans 134, 137, *85*, *86*, *87*, *88*, *89*
Transportation maps 125, 127, 128, 129, 131, 132, *82*, *83*, *84*
Triangular distance table *79*
Triangulation 53, 54, *30*
Ts'ai Lun 61
Turgot, Michel Etienne, plan of Paris 143, 145, *94*
TWA Atlantic River map *119*

Underground map, of London system 108, 129
Universal Transverse Mercator Grid System (UTM) 31, *16*
US Coast and Geodetic Survey 55, 115, 183
US Commission on National Atlases 102
US Geological Survey 25, 27, 42, 180, 183, *115*, *116*
US Public Land Survey 101
US Naval Hydrographic Offie 115
Uses of maps 11, 15–18, 23
Utopia, map of *7*

Vallard, Nicolas 108
Vandermaelen, Philippe Marie Guillaume 129
van Deutecum, Baptista 44, 109
van Deventer, Jacob 93, 139
van Keulen, Johannes 109
van Langeren, Jacob 29
Vauban, Sebastian de 143
Vavassore, G. A. 108
VEB Hermann Haack 187

Vellum 61
Vesconte, Petrus 104
Vineyards, symbol for 47
Visscher, Claes Jansz 145
Vocabulary of cartography 39–51
von Breydenbach, Bernhard 120
von Humboldt, Alexander 161, 168

Waghenaer, Lucas Janszoon 109, *72*
Wainwright, Alfred 194
Waldseemüller, Martin 27, 43, 63, 82, 83
Walker, Francis, Amasa 167
Washington, George 101
Watermark 61
Waterway maps 127, 128, *82*
Wax engraving, or cerography 68
Waywiser 55

Webb, Montague *124*
Welser, Marcus 119
Westermann Verlag 186
Wheel maps 74
Whewell, William 159
White, John, map of Virginia *66*
Wine maps 191, *120*
Woodcock, John 191
Woodcut maps and plans 40, 44, 48, 61–4, *27*, *37*
World Aeronautical Chart (WAC) 131, 170
Wright, Edward 108

Yates, William, map of Lancashire *30*
Young, Arthur 167

al-Zargali 73